THE OPEN UNIVERSITY
A SCIENCE FOUNDATION COURSE

UNIT 1 SCIENCE AND THE PLANET EARTH

THE OPEN UNIVERSITY

THE SCIENCE FOUNDATION COURSE TEAM

Steve Best (Illustrator)
Geoff Brown (Earth Sciences)
Jim Burge (BBC)
Neil Chalmers (Biology)
Bob Cordell (Biology, General Editor)
Pauline Corfield (Assessment Group and Summer School Group)
Debbie Crouch (Designer)
Dee Edwards (Earth Sciences; S101 Evaluation)
Graham Farmelo (Chairman)
John Greenwood (Librarian)
Mike Gunton (BBC)
Charles Harding (Chemistry)
Robin Harding (Biology)
Nigel Harris (Earth Sciences, General Editor)
Linda Hodgkinson (Course Coordinator)
David Jackson (BBC)
David Johnson (Chemistry, General Editor)
Tony Jolly (BBC, Series Producer)
Ken Kirby (BBC)
Perry Morley (Editor)
Peter Morrod (Chemistry)
Pam Owen (Illustrator)
Rissa de la Paz (BBC)
Julia Powell (Editor)
David Roberts (Chemistry)
David Robinson (Biology)
Shelagh Ross (Physics, General Editor)
Dick Sharp (Editor)
Ted Smith (BBC)
Margaret Swithenby (Editor)
Nick Watson (BBC)
Dave Williams (Earth Sciences)
Geoff Yarwood (Earth Sciences)

Consultants:
Keith Hodgkinson (Physics)
Judith Metcalfe (Biology)
Pat Murphy (Biology)
Irene Ridge (Biology)
Jonathan Silvertown (Biology)

External assessor: F. J. Vine FRS

Others whose S101 contribution has been of considerable value in the preparation of S102:

Stuart Freake (Physics)
Anna Furth (Biology)
Stephen Hurry (Biology)
Jane Nelson (Chemistry)
Mike Pentz (Chairman and General Editor, S101)
Milo Shott (Physics)
Russell Stannard (Physics)
Steve Swithenby (Physics)
Peggy Varley (Biology)
Kiki Warr (Chemistry)
Chris Wilson (Earth Sciences)

The drawing on the front cover is from Copernicus's original manuscript, and places the Sun at the centre of the Universe.

The Open University, Walton Hall, Milton Keynes, MK7 6AA

First published 1987. Reprinted 1989, 1990, 1992, 1993, 1994, 1995.

Designed by the Graphic Design Group of the Open University.

Filmset by Santype International Limited, Salisbury, Wiltshire; printed by Henry Ling Ltd., at the Dorset Press, Dorchester, Dorset.

ISBN 0 335 16325 4

This text forms part of an Open University Course. For general availability of supporting material referred to in this text please write to: Open University Educational Enterprises Limited, 12 Cofferidge Close, Stony Stratford, Milton Keynes, MK11 1BY, Great Britain.

Further information on Open University Courses may be obtained from the Admissions Office, The Open University, P.O. Box 48, Walton Hall, Milton Keynes, MK7 6AB.

1.7

STUDY GUIDE

This Unit consists of three components: the text, the associated experiments, and the TV programme 'The planet Earth — a scientific model'. You will find notes relating to this TV programme at the end of the text.

We are conscious that you may not have studied for some time. So, as this is the *first* Unit of the Course, we shall take up somewhat more space with this Study Guide than will be normal in later Units, in order to offer some specific advice on how to tackle this and other Units. More general advice on how to study is given in the *Introduction and Guide*, which you should by now have read carefully.

The text You should read the text thoroughly — it is the core of the Unit. In order to help you to assimilate this material, we have provided plenty of questions for you to answer along the way. There are two formal types of question — In-Text Questions (ITQs), and Self-Assessment Questions (SAQs). ITQs are specifically designed to carry the argument forward, so you should always do such questions *when you come to them*, and then refer to their answers before you move on. SAQs, on the other hand, are designed to help you to consolidate ideas that you have already met; they enable you to assess for yourself whether you have understood the main terms and concepts introduced in the Unit, and whether you have mastered the particular skills or techniques listed in the Objectives for the Unit. These Objectives, together with the answers for both the ITQs and SAQs, are to be found at the end of the Unit.

Whereas most of the Unit is about the relative motion of the Sun, Earth and Moon, Section 1 comments briefly on the nature of science and its methods. Certain important terms, such as model, that recur throughout the Course are introduced here. This material will have its proper impact only when you have become familiar with some concrete science which illustrates the general points that it is trying to make. We recommend you to read quite quickly to the end of Section 1.2.1 and then, when you have read all the way through the Unit, to reread Sections 1.2 and 1.2.1 in conjunction with the concluding remarks of Section 6.

Experiments Two simple but important experiments are integrated into the Unit in Sections 3.5.2 and 4.1.2. To do them, you will need: three or four metres of thin string; two small heavy objects that can be hung from the string (e.g. good-sized bunches of keys); and a watch that can measure seconds. When you come to these experiments, you may think you know the result you will obtain; but be careful, you may well be in for some surprises. You should certainly not be tempted to skip these experiments, for they are carefully designed to carry the argument of the text forward; omitting them may well hinder your understanding of the subsequent text. Also, they are an important component of the week's work and may well be assessed in the associated Computer-Marked Assignment (CMA).

TV programme, 'The planet Earth — a scientific model' This programme is designed to reinforce your understanding of several of the ideas presented in the text. In particular, it discusses the following in more detail: the Foucault pendulum experiment and its contribution towards our knowledge of the Earth's spin; the phases of the Moon and how they relate to the motion of the Moon about the Earth; and the apparent reversals in the progression of the planets across the sky. In order to derive maximum benefit from this programme you should already have read through the text at least once before the programme is transmitted. (Note, however, that this is *not* a general rule. It will sometimes be possible, and even desirable, for you to watch a TV programme before, or perhaps during, your first reading of the associated text.) You will probably find it helpful to read through the introductory paragraph of the TV Notes (Section 5) *before* watching the programme; it will give you an indication of what to expect. Read through the rest of the notes when you have watched the programme, to help you consolidate what you have learnt.

If you complete this Unit ahead of schedule (well, all things are possible!), we suggest that you move straight on to Unit 2. There is a logical continuity between the Units that makes this the natural thing to do. But there is also the point that, whereas Unit 1 concentrates on qualitative concepts and the visualization of scientific models, Unit 2 begins to ask *quantitative* questions, such as 'how big?' and 'how far?'. If your mathematics is at all rusty, you should try to allow yourself as much time for Unit 2 as possible.

Finally, if you have not already done so, we suggest that you now obtain as many items as you can from the shopping list in the *Introduction and Guide*; this will save much time and inconvenience later in the Course when you come to do the various experiments. In addition to the items that you will need for this Unit's experiments (see above), the most urgent things to obtain are the ones you will need for Units 2 and 3:

a piece of straight dowelling at least 1.2 metres long (a broom handle would do nicely!);

a tape-measure or rule (with centimetre and millimetre markings) at least 1.5 metres long;

some Blu-Tack or plasticine;

a pair of compasses and a sharp pencil;

a protractor.

1 INTRODUCTION: SCIENCE IN PERSPECTIVE

'The whole of science is nothing more than a refinement of everyday thinking.'
(Einstein, *Physics and Reality*, 1936)

You might expect a science foundation course to start with a simple definition of science, but this is not easily done. The problem is that science is a very complicated business, and it can be looked at from a variety of angles. When people look at something complicated from different directions, they see different things. So all that we shall do at first is to sketch the view from two popular standpoints. As you work your way through the Course, you may find that these perspectives are a useful way of framing what you have learnt.

Look at the photograph of the planet Earth in Figure 1. It was taken in 1968 from a satellite at a range of about 38 000 km (24 000 miles). When they look at such a photograph, some people think first of its cost, and ask whether it was worth all the money and effort of sending up a satellite. Here, science is viewed from a *social* standpoint. The photograph was not the work of an isolated individual, but of a complex industrial society with considerable technical skills. The satellite from which the photograph was taken was the work of many hands. The calculations that put it in the right place had been developed by generations of scientists. And finally, the decision to undertake the enterprise was a political one: it involved the postponement of other social and scientific projects which were assigned a lower priority.

Other people might find this photograph attractive because of its romantic associations with space travel. Few people were not moved by the thought of fellow human beings drifting about in space and (later) walking on the Moon. There is no doubt that human curiosity is the most elemental, most basic driving force behind science. For many of us, questions such as 'What does the dark side of the Moon look like?' or 'What is it like on the Moon's surface?' have an irresistible appeal. In this particular case, the photograph is, for many people, direct proof that the Earth is round. Seeing, it seems, is

FIGURE 1 The planet Earth. A black-and-white copy of a NASA photograph taken on 21 January 1968 from satellite ATS 3.

believing. Yet it has been known for centuries that the Earth is round, although nobody could actually *see* it until space travel became possible. The evidence of direct observation has a more powerful impact than conclusions reached by reasoning.

These feelings of curiosity and different degrees of conviction are the experiences of individuals, and they differ from the social matters discussed earlier. They are concerned with how we get to know things. Questions of this sort are raised when we examine the methods of scientific inquiry, the ways in which scientists make their discoveries.

The aim of science, then, is to understand the working of the natural world about us. We shall now comment further on science as a social activity, and then discuss how science is practised.

1.1 SCIENCE AS A SOCIAL ACTIVITY

When we speak of science as a social activity, the society that we first think of is the whole complex of social, economic and industrial relationships within which we live. Science influences the structure and the demands of this society, and society returns the compliment. An example of this mutual interaction is the development of nuclear energy. Nuclear fission — the breaking up of the tiny nuclei that are at the cores of atoms — was discovered almost accidentally by Otto Hahn and Fritz Strassman in 1938. This scientific finding very quickly had a profound effect on society. The perspectives and conduct of prominent social activities such as war and electricity production were radically changed by the development of nuclear weapons and nuclear power-stations. To obtain these things, society had to make demands on scientists, and to allocate the necessary resources. Consequently, both during and after World War 2, the work of thousands of scientists was focused on nuclear physics, and in particular on the study of uranium, plutonium and related chemical substances. This had a very important influence on the directions taken by post-war physics and chemistry.

An important intermediary in the mutual influence of science and society is technology. Science first raised the prospect of nuclear energy; but the practical implementation of it called for high-grade engineering skills. For example, new methods had to be designed to handle, machine and produce novel metals and alloys on a large scale. This illustrates how the development of science and technology in modern industrial societies is influenced by political and economic decisions on expenditure and investment. It is always worth considering whether the money is being wisely spent.

FALSIFIABILITY CRITERION

Now let us turn to science in the context of a more modest society: the society of scientists themselves. Science after all is the outcome of the combined efforts of this social group. Many interesting matters might be raised about the way the group functions: how they referee one another's publications, publishing some and rejecting others; how they set up committees; and how such committees dispense research money and award Nobel Prizes. However, here we shall concentrate on a common moral assumption that the members of the group share: when scientists read of one another's experiments or calculations, they assume that they are reading a truthful account, written in good faith. This mutual trust allows scientists to benefit from one another's work. Without it, the development of science would have been abysmally slow.

The importance of the assumption of truthfulness is shown by scientific attempts to investigate 'paranormal' phenomena such as telepathy or spoon-bending (Figure 2). It is not that there is nothing of scientific interest here — the problem is that the investigator and the investigated *may* not be engaged in the same enterprise. One is a scientist anxious to find out something new about the natural world; the other may be a magician or conjurer anxious to make a living out of highly developed skills of deception. The difficulty of ensuring a shared commitment is one reason why paranormal phenomena have had so little influence upon science.

Thus scientists have to rely upon the written reports of other members of the society of scientists. This explains why a scientist caught lying in print is treated much more severely than, say, a journalist. In the 1970s, the scientific reputation of the eminent psychologist, Sir Cyril Burt, plummeted following the publication of evidence that he had fabricated results on the intelligence testing of separately reared identical twins. He invented results which supported his belief that intelligence is largely hereditary. If caught in such acts of fraud, the scientist will fall from grace very rapidly indeed.

FIGURE 2 A spoon-bender.

1.2 THE SCIENTIFIC METHOD

How is scientific knowledge obtained? What is distinctive about the scientific method? Few scientists give much thought to these questions, but among philosophers they are very controversial. What we shall do here is to put forward a simple version of one common theory of the scientific method (a theory associated with the name of Sir Karl Popper), and then express a few reservations about it.

Everyday experience teaches us a good deal about the world around us. In particular, certain patterns and regularities soon become apparent. For example, consider an observer who, while he was looking out of his study window today, noticed that the Sun rose on his left and set on his right. The observer can make a formal record of this:

1 'Today, 11 November 1986, when I faced the oak tree opposite my study window, the Sun rose on my left, and set on my right.'

Suppose also that the observer recalls that the same thing happened yesterday:

2 'Yesterday, 10 November 1986, when I faced the oak tree opposite my study window, the Sun rose on my left, and set on my right.'

Because of this curious coincidence, the observer may now hazard a third statement:

3 'If I stand facing the oak tree opposite my study window, the Sun always rises on my left, and sets on my right.'

Statements 1 and 2 are *singular* statements which describe a state of affairs at one particular place and time. Statement 3, although prompted by statements 1 and 2, is clearly different in kind. In particular, it is much more *general*: it claims to tell us what has always happened in the past, and what will always happen in the future.

The process of generating general statements from singular statements is called *induction*. It is an essential part of science, yet it is not what distinguishes science from other fields of study. Induction is an exercise of the *imagination*: it leads from what *is*, to what *might be*. A stage in which the imagination works on what is already known is something that science shares with politics, religion and the arts. What is distinctive about science is the *form* of its general statements: in particular, they can be *tested* by observations and experiments. This makes such statements potentially falsifiable. According to the **falsifiability criterion**, a general statement can be scientific if and only if one can conceive of an observation or experimental result that can logically prove it wrong.

Consider statement 3 again. Its virtue lies in its power of prediction. If it is true, then the observer can derive from it a new singular statement:

4 'Tomorrow, 12 November 1986, if I face the oak tree opposite my study window, the Sun will rise on my left and set on my right.'

This derivation of a singular statement from a general statement is a process known as *deduction*. In its direction, it is the opposite of induction: from the general to the singular rather than vice versa. But the difference goes deeper than this. Deduction is a *logical* process: if statement 3 is true, then statement 4 *must also be true*. By contrast, induction is an imaginative process, not a logical one. When a general statement is created from a collection of singular statements, as with statement 3, we cannot be sure that it is true: it may well be false, and only time and experience will tell.

This brings us back to the point that statement 3 is potentially falsifiable. From it, the singular statement 4 has been deduced, and this can be tested by experiment. Suppose that tomorrow the observer witnesses sunrise and sunset from his study window. If the Sun rises to the left and sets to the right, then statement 4 is true, and statement 3 has survived a test. But if, for instance, the Sun rises to the right and sets to the left, then it is not just statement 4 that is false: statement 3 is false as well, and must be rejected.

Notice the asymmetry of these two possible outcomes. If statement 4 turns out to be false, statement 3 is false as well. But if statement 4 turns out to be true, it does not follow that statement 3 is also true: it remains possible that the Sun will rise and set in unexpected places the day after tomorrow. We can never say that the general statements of science are true; when they survive experimental tests, we say rather that they have been supported or *corroborated* by new evidence. If statement 4 turns out to be true, we retain statement 3 simply because it has not yet been falsified. And if retained, then like other general statements of science it could be useful. For example,

MODEL

suppose the observer works all day, and therefore does his washing and hangs it out to dry at night. He can now change the position of his washing line so that it catches the morning Sun.

One important characteristic of statements judged to be non-scientific by the falsifiability criterion, is that they often contain let-outs which prevent falsification. For example, consider the statement:

The Sun moves across the sky because it is pushed by an angel whose presence cannot be detected.

Here the let-out lies in the words 'cannot be detected': falsification is impossible, by definition.

Our trivial preoccupation with the view from the observer's window may seem far removed from real science. Recall, however, Einstein's words quoted at the beginning of this Unit: 'The whole of science is nothing more than a refinement of everyday thinking.' Many important scientific discoveries have followed the kind of sequence we have just described. Contact with a set of observations triggers the formulation of a general statement. A general statement in science may be called a law, a hypothesis or a theory, a distinction that we shall not elaborate here. What is important about the general statements is that new, singular statements or predictions can be drawn out of them by deduction. These predictions can then be tested, for example by experiments that have not previously been tried. If the predictions turn out to be wrong, the theory can be regarded as falsified, and then rejected. But if they are correct, the new theory may arouse the interest of other scientists and gain their provisional acceptance. Indeed, a truly great theory will, by its boldness and imagination, excite new thoughts, and also give rise to tests and experiments that would otherwise not have been done. In such a case, the theory can be described as fertile: it creates a large, new and complex research programme of its own. Later in the Course, you will see how the theory of continental drift did this in the Earth sciences (Units 7–8), and how Mendeléev's idea about the behaviour of the chemical elements, the Periodic Law, achieved it in chemistry (Units 13–14).

What has been described above certainly captures important aspects of science, notably the inductive–deductive sequence, and the testing of the deduced consequences of laws and theories by experiment. However, its emphasis on falsification has aroused a good deal of argument. First, there is the question of whether all scientific theories are falsifiable. This has been raised in connection with the theory of evolution which you will meet in Unit 19. The theory can certainly be tested and corroborating evidence for it can be found, but it is hard to imagine experiments that could falsify it. Second, a single falsifying experiment rarely has the cutting edge needed to destroy an important theory, especially if the theory is well established and there is nothing to put in its place. Falsification takes time, mainly because scientists are reluctant to abandon the theories through which they have learnt their trade and made their living. And finally this view of science stresses a rather particular role of experiments: they are attempts to falsify or, failing that, to corroborate laws and theories. But modern experiments often call for sophisticated equipment and great technical skills. The sheer joy of exercising such skills in unknown scientific territory will often carry a scientist through an experimental programme without a thought of the possible falsification of a theory. Indeed, we hope that you will catch a flavour of this in Unit 2 when you measure the distance between the Earth and the Moon.

Yet whatever one's reservations, the vision of science discussed in this Section is an exhilarating one: experience combines with imagination to produce bold and exciting theories. But then, admitting the possibility of error, science tries to destroy the fruits of its own imagination with experiments. Either error is exposed by falsification, or the theory survives — but survives only provisionally. For science can never give us *final* certainty or truth; it advances over the wreckage of its own rejected theories. Is this really science, or is it a prescription for what science ought to be? When you have finished this Course, you will be better placed to decide.

1.2.1 MODELLING: A PART OF THE LANGUAGE OF SCIENCE

When you describe things to other people, you often use metaphors or analogies. Thus, 'my brother is a snake' or 'the Chairman of the S102 Course Team is a saint'. Scientific theories also contain metaphors, but they are normally more elaborate than those of ordinary speech. Scientists call them *models*. The term can be defined as follows:

> A **model** is an artificial construction invented to represent or to simulate the properties, the behaviour or the relationship between individual parts of the real entity being studied.

This definition may appear somewhat abstract, so here are some concrete examples:

Example 1 The human heart is a complicated organ, but its main role may be thought of as that of a pump. Thus a pump in a central heating system is a model of the heart.

Example 2 Long before space travel, certain observations led people to conclude that the Earth was round. Thus a sphere became a model for the Earth.

The next example comes from Unit 20. Read it carefully once, and do not worry if you do not fully understand it. You should still be able to follow the subsequent comments.

Example 3 The controlled interbreeding of tall and short pea plants yields only tall plants in the first generation of offspring, but an approximately three-to-one ratio of tall to short plants in the second. This can be understood by using a model of inheritance. Each mating plant contributes something represented by one of a pair of letters, either T or t, to form a new pair, which can be either TT, Tt or tt. The combinations TT and Tt yield tall plants; the combination tt yields short plants. The original tall plants are each represented by the pair TT, and the original short plants by the pair tt. In each generation thereafter, the relative proportions of the possible combinations of letters then account for the ratio of tall to short plants.

Notice first that when we use the term model, we are not speaking of plastic miniatures of ships or aeroplanes. In none of examples 1–3 does the word mean a scaled-down replica of a real thing. What we have instead is a situation in which scientists single out for attention some narrow aspect or particular property of a more complicated system. To help their thinking, they construct a model. The model represents the system freed of the additional complications that the scientist has chosen to ignore; it has been deliberately 'tooled' to do the particular job in hand. Thus in both examples 1 and 2 the complicated internal structures of the heart and the Earth do not feature in the model, and in example 2 the fact that the Earth is not exactly spherical is ignored. The model described in example 3 is even more striking in that it reduces the variations in size of a number of pea plants to various combinations of two letters representing the particular character under investigation. A shuffling of these pairs of letters, to form new pairs, enables us to predict the relative proportions of tall and short plants in a series of generations.

Example 3 also provides an instance of a particularly important type of modelling: *mathematical* modelling. In this particular case, the mechanism of inheritance is regarded as a mathematical process whose consequences can be calculated by the well-established laws of chance and probability.

In these models, there are signs of the intellectual ruthlessness of scientists. The wholeness of things is ignored or brushed aside in the impatient search for precise explanations of some particularity. Some great artists have found this aspect of science repulsive. One catches something of this in William Blake's portrayal of Isaac Newton (Figure 3) and in Jonathan Swift's account of the Laputians in *Gulliver's Travels*. The Laputians are

STAR

FIGURE 3 'Newton', by William Blake (1757–1827), the English painter, poet, engraver and mystic.

obsessed with mathematical problems and astronomy to the exclusion of other human and practical concerns. Gulliver's description of them symbolizes their two obsessions:

> Their Heads were all reclined either to the Right, or the Left; one of their Eyes turned inward, and the other directly up to the Zenith.

Nor are their actions any more reassuring than their appearance. The typical Laputian:

> is always so wrapped up in Cogitation, that he is in manifest Danger of falling down every Precipice, and bouncing his Head against every Post.

But do not be intimidated by this propaganda. Even if there is something in it, the study of science offers considerable consolation! You can now start to find out just how much.

SUMMARY OF SECTION 1

1 Science is a social activity. It influences the structure of society, and society's demands influence the structure of science.

2 If scientists could not depend upon the truthfulness of other scientists' reports, rapid progress in science would become impossible.

3 According to one school of philosophy, science involves the creation of general statements or theories from observations and experience. This is an imaginative process. The general statements of science are falsifiable: they can, in principle, subsequently be refuted by testing their deduced consequences through experiments or further experience.

4 Models are an important part of the language of scientific theories. They are intellectual constructions, created by removing from the original all factors that appear to be of no relevance to the immediate scientific problem.

The following Self-Assessment Question, like all SAQs, may be attempted when you come to it in the text, and/or when you have completed the Unit.

SAQ 1 Assuming that scientific statements are those that conform to the falsifiability criterion of Section 1.2, try to classify the following as either scientific or not scientific.

(a) All ravens are black.

(b) If two bodies at different temperatures are placed in contact with one another and are isolated from their surroundings, the colder body never warms up.

(c) Somewhere in the Universe, there exist, or have existed, extra-terrestrial life forms.

(d) At the poles of the planet Mars, the planetary surface is always crawling with bug-eyed monsters.

(e) One day, capitalism will everywhere be overthrown.

(f) There is a ceremony that, if it is correctly performed, conjures up the Devil.

(Answers to SAQs are given at the back of the Unit, starting on p. 48.)

2 WHAT SHAPE IS THE EARTH?

In the rest of this Unit, you will be asked to consider questions concerning the Earth and its relationship to the Sun and the Moon. You probably already know quite a lot about this relationship from general knowledge; even if you do, bear with us for a moment.

How did you come to know that the Earth is round and that it spins about its axis and orbits around the Sun? More likely than not, you have just been led to believe it, because your parents and your teachers told you it was so from your early days. But can you *justify* why you believe it? Your eyes will tell you that the surface of the Earth is flat or undulating, depending on where you are, but certainly not that it is spherical. And you cannot possibly have any direct evidence from your senses about the *motion* of the Earth — when you keep still, the Earth seems to keep still as well. What you have accepted is a *model*, worked out by previous generations of scientists and gradually adopted into the pool of human knowledge.

It is a worthwhile exercise to try to forget all this knowledge for a moment and to retrace its derivation from everyday experience. The purpose of the remainder of this Unit is to guide you through this derivation, and to illustrate from this example how different aspects of scientific reasoning work in practice and how theoretical models are constructed, modified or rejected.

2.1 THE PANCAKE MODEL

Imagine for a moment that you are a member of a tribe living in the very distant past, in territory that consists of a large and relatively flat area. The area comprises meadows, lakes and woods, and is surrounded by impenetrable forests. Your tribal community supports itself entirely by hunting, fishing and primitive farming.

What sort of a picture could your tribe have about the place where they live? Surely they would be aware of the Sun, and of the fact that it provides warmth and also light that makes wild beasts visible for hunting (or avoiding). They might even notice that the heat and the light from the Sun help plants to grow (if only because they do not grow well in places shielded from it). On the other hand, they will also observe the Moon and the **stars**, which have no obvious relevance to everyday life. (You can think of a star as one of the vast number of incandescent celestial objects that can be seen in the night sky; this is certainly not a rigorous definition, but is adequate for the purposes of this Unit.) Even less understandable are some other natural phenomena such as wind, rain, snow, thunderstorms and earthquakes. They do not exhibit any obvious regularities and their effects on the well-being of the tribe are unpredictable, sometimes good, sometimes bad.

It is reasonable to speculate that an intelligent member of the tribe could formulate the following model of the world: the place where the tribe lives (the Earth) is like a flat pancake, and it does not move (earthquakes excepted); and the Sun is some sort of a glowing disc or fireball that moves repeatedly over the pancake. All other detail — such as what happens to the fireball when it disappears, how it gets over to the other side of the pancake, whether it is indeed the same fireball all the time or a new one every day — would be quite beyond this person's grasp.

Yet, in spite of its crudity, this very primitive pancake model already has an element of scientific approach in it. It is not just a direct, raw experience, it is an artificial mental image of something that cannot be directly perceived as a whole.

How would this model of the Earth be modified by another tribe that lived by the seashore or on a island, instead of being land-locked?

Well, their range of direct observations would be wider because they would observe the surface of the water extending as far as they could see, and would probably notice regular tides and possibly even the connection between the tides and the Moon. They would still regard the Earth as a pancake or a disc, but it would be either surrounded by water or floating on water like a huge boat. They might still regard the Earth as being 'at rest' (which is jargon for 'stationary'), because there is no sensation of its motion, not even of its floating on the water. The Sun would be moving around the Earth–water disc, or alternatively emerging from the waters and descending into them. The stars in the sky would be no more than some very distant and unreachable stage-setting; but the Moon could possibly be recognized as some object or mysterious being that has some influence on the tides, and that moves across the sky in much the same way as the Sun.

As long as direct observational experience is limited to a small part of the Earth's surface, the pancake model is the most sensible one, indeed perhaps the only reasonable one. There is nothing to suggest otherwise.

2.2 THE SPHERICAL MODEL

The flat-disc or pancake model of the Earth had to be abandoned only when people were able to collect and compare observations from much larger areas of the Earth's surface. These observations provided two kinds of evidence to suggest that the surface of the ocean is curved, not flat.

First, a large departing ship appears to be sinking into the waters on the horizon (Figure 4b), and an arriving ship to be emerging from them. Yet, sailors know that their ships remain on the surface all the time, and they in turn will testify that they saw the harbour and the shore sinking or emerging.

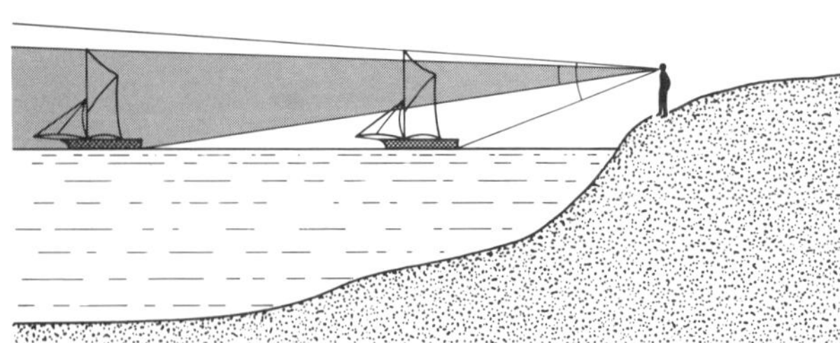

(a) flat Earth

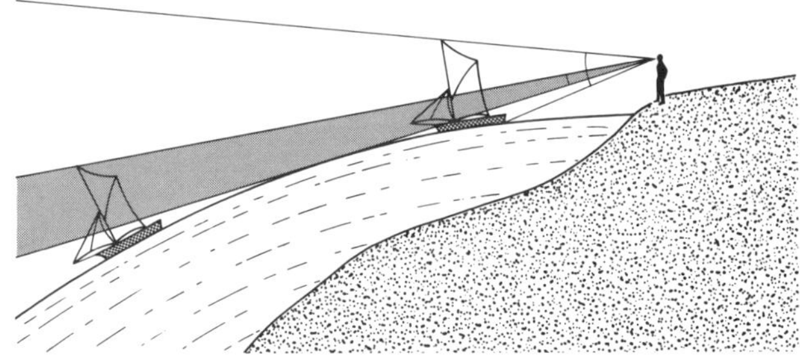

(b) curved Earth

FIGURE 4 Visual evidence that the Earth is not flat. (a) If the Earth were flat, a departing ship would be seen to be diminishing, but always complete.

(b) A departing ship actually appears to be sinking as it disappears below the observer's horizon.

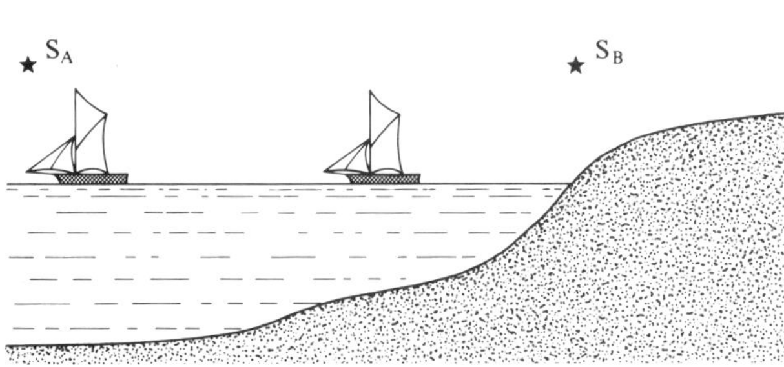

(a) flat Earth

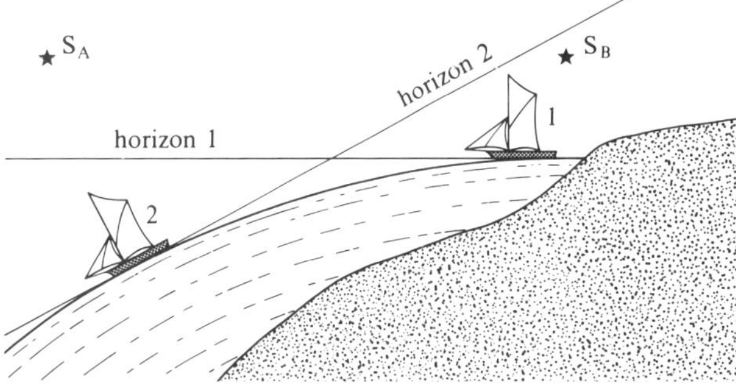

(b) curved Earth

FIGURE 5 We know from observation of the stars that the Earth is not flat. (a) If the Earth were flat, both stars S_A and S_B would be seen from the ship, no matter how far it travelled. (b) When the ship moves from position 1 to position 2, star S_A moves higher above the horizon and star S_B disappears.

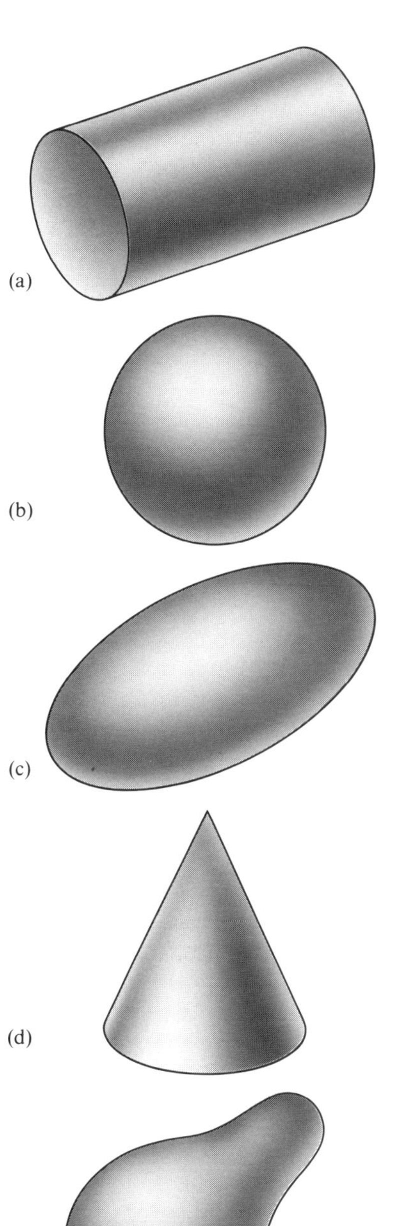

FIGURE 6 Five kinds of curved surface: (a) cylindrical; (b) spherical; (c) ellipsoidal; (d) conical; (e) pear-shaped.

The only possible explanation of this observation is that the surface of the water is not flat. And if the ocean surface is curved, it could be argued plausibly that the surface of the Earth *as a whole* is curved. (By the way, did you notice that this explanation involves the hidden assumption that everybody involved in these observations sees 'straight' or, in other words, that light travels in a straight line? This is a reasonable assumption in this case, but it is certainly not absolutely true!)

Second, the positions of the stars in the sky start to change as a ship travels south or north. The stars on one side of the horizon appear to the voyagers to be attaining higher and higher altitudes, and the stars on the other side appear to be moving lower down or even disappearing altogether. On the return journey, these changes are gradually reversed until all is the same again on return to the home port. It is obvious that these changes are not *real* changes in the appearance of the sky, because those who do not travel do not observe them. The changing pattern in the motion of the stars must therefore be due to the *changing position of the observer* in the ship. If the Earth and the ocean surface were essentially flat, this would be impossible. Provided that the distance to the stars is much larger than the distance travelled on Earth, then all stars would be seen in the same positions from all places on the land or the sea. Once again, the only possible explanation of the observed changes lies in accepting that the surface of the ocean is not flat, but curved (Figure 5).

Now try the following In-Text Question (ITQ). Remember that you should try to do the ITQs as soon as you reach them in the text, and then check your anwers at the back of the Unit (in Unit 1, the ITQ answers start on p. 46). ITQs are an integral part of the teaching, so don't put off doing them till later!

ITQ 1 (a) On the basis of a single observation, of either of the kinds illustrated in Figures 4 and 5, onc can conclude only that the Earth's surface is curved in *one* direction, namely along the direction in which the ship is moving. Can you tell, on this evidence alone, whether the surface is cylindrical, spherical, ellipsoidal, conical, pear-shaped or of any other curved shape (see Figure 6)?

(b) If not, what additional evidence would you need in order to show that the surface is spherical?

In doing ITQ 1, you have seen that although direct visual observation provides evidence that the surface of the ocean is not flat, it is not so easy to see why it should be spherical. True, it is possible in principle to repeat the observations and measurements from different places and in different directions, as we suggested in the answer to ITQ 1. However, all anyone could deduce from such measurements would be that it is *not impossible* that the Earth is spherical or, at the most, that it is *likely* to be spherical. But such a statement would be highly unreliable, particularly considering the state of technology at the time when a spherical model of the Earth was first suggested (about 500 BC, on present evidence).

GEOCENTRIC MODEL OF THE UNIVERSE

UNIVERSE

CULMINATION POINT

It is instructive to follow the reasoning that led the Ancient Greek philosopher Pythagoras to suggest that the Earth *is* spherical. He was aware of the curvature of the part of the Earth that he knew had been explored. He also attached great significance to the fact that the Sun and the Moon were either circular discs or spherical objects. Stars in the sky appeared to follow circular paths during the night. Experience told him that, of all possible curves, the circle looked the most regular and was the most easily drawn, that a circular cross-section was the most effective shape for the wheels of carts, and that spherical objects could be rolled in any direction more easily than objects of other shapes. Thus he formulated the idea that a circle is the most perfect curve and that *a sphere is the most perfect shape* that any object in the world can have. And because the Earth to him was something of prime importance and perfection, at least as important and perfect as the Sun or the Moon, it seemed only natural to believe that it *must* be spherical.

This model of a spherical Earth represents a great advance of scientific thinking over the pancake model. First, it takes into account the observation that the surface of the ocean (and therefore probably of the whole Earth) is not flat. Second, it attributes to the Earth a shape that is known to be the shape of other important and 'perfect' things. Last but not least, it offers a very simple solution to the problem of what happens to the Sun after it has set. There is no need to worry any more about the possibility that it disappears into the ocean and is reborn again in some mysterious way the next morning. For, if the Earth is a sphere, then it somehow seems much more reasonable to argue that the Sun (as well as the Moon and the stars) travels in a circle around the Earth.

Although Pythagoras was the first (as far as our records show) to propose that the Earth has a spherical shape, it is the Greek astronomer Ptolemy who is now associated with the so-called **geocentric model of the Universe** (note that the **Universe** is the totality of everything that exists, not just the comparatively tiny region around the Earth and the Sun).

In about AD 150, Ptolemy produced an extensive catalogue of celestial objects and explained their motion in the sky by assuming that the Earth is a stationary sphere at the centre of the Universe. (Hence the name geocentric, where the prefix 'geo' is the Greek for Earth.) According to this model, the Sun, the Moon, the stars and the planets all travel around the Earth, either in simple circles, or in more complicated paths that can nevertheless be explained by combining one main circular motion with additional motions in much smaller circles around the main path.

The geocentric model, so attractive in its simplicity, also conformed with the Christian teaching concerning the creation of the world and humankind, and concerning the special role of the Earth in the Universe. It is interesting that there was a different yet parallel stream in Greek philosophy (associated with Philolaus and Aristarchus) that regarded the Sun, not the Earth, as the centre of the Universe. Although this alternative model had at least as much merit as the geocentric model on the evidence then available, it was widely rejected on philosophical grounds.

SUMMARY OF SECTION 2

In this Section you have read about some of the observations, assumptions and reasoning that have led us to adopt a spherical model for the shape of the Earth. Also, you have met the geocentric (Earth-centred) model of the Universe. More important, you have seen the scientific method in action: observations gave rise to ideas, which could eventually be turned into a pancake model of the Earth's shape. But when further observations, coupled with reasoning, falsified this model, it was abandoned in favour of another model (the spherical model) that gave better agreement with the observations. Notice how well this sequence fits in with the account of the scientific method and of modelling given in Section 1.

Note that the spherical model was adopted long before there was any direct visual evidence of the spherical shape of the Earth. Yet scientists had faith in it because it was consistent with other observations that it alone could satisfactorily explain. Science is not just about what we can see; it also uses reasoning to ensure that all the relevant observations and evidence are consistent with one another and with the theory under examination.

3 THE EARTH AND THE UNIVERSE: EVERYDAY EXPERIENCE AND OBSERVATIONS

In the previous Section, you were mainly concerned with the Earth and the ways in which ideas about its shape were deduced from everyday observations. The models suggested depended on the life-style and natural environment of those making the observations. The relationship between the Earth and other celestial bodies, in particular the Sun and the Moon, were mentioned only briefly.

In theory, the simplest and most reliable answer to the question of how the Earth relates to the Sun and the Moon would be found by travelling sufficiently far away from all three of them and having a look 'from the outside', as it were. Alas, this is not yet fully possible for human beings. Nevertheless, we should still be able to develop an understanding of this relationship, even within the limitations of our Earth-bound observations. What we must do is to analyse these observations carefully, and use the information that we glean to build a satisfactory picture of the relative positions and motions of all three bodies.

In this Section, we shall summarize the available evidence. Then, in the next Section, we shall develop a simple model based upon this evidence.

3.1 DAYS AND NIGHTS

The regular sequence of days and nights is one of the most common experiences shared by all people, and indeed by all living creatures on the Earth. It is taken so much for granted that it represents an ideal of certainty, permanence and regularity.

Yet a closer look soon reveals that this sequence does not have a perfectly rigid pattern. To begin with, it is *not the same for the whole Earth*. The Sun rises earlier for an observer in Tokyo than for one in Paris. Nowadays this can be easily established, and the difference in time accurately measured, thanks to long-distance communications by telephone or radio. Moreover, even an observer who stays put in one place will experience *variations in the relative length of daylight and darkness*. These variations are related to the position of the Sun in the sky. Approximately half-way through the daylight period (noon), the Sun reaches its highest point above the horizon, known as the **culmination point**. This culmination point does not remain at the same place all the time, and it is common knowledge that days are long when the Sun culminates high above the horizon, and short when the culmination point is low. The relationship between the height of culmination and the length of daylight connects the day–night cycle closely with the cycle of seasons.

Before we delve further into this connection, let us emphasize one crucial observation about the day–night cycle. Although there are local as well as seasonal variations in the relative length of days and nights, the time interval *between two consecutive culminations* of the Sun is constant. It is the

SOLAR DAY

LUNAR PHASES

same for all observers anywhere on the Earth and is the same throughout the year (except for the extreme polar areas, where the Sun does not appear above the horizon for a long time during polar winters). For this reason, the interval between two consecutive culminations of the Sun became one of the first natural units for measuring time, and is known as a **solar day**.

3.2 THE FOUR SEASONS

When the relative durations of daylight and darkness within each solar day are systematically measured and recorded over several years, a clear pattern emerges that repeats itself regularly. This pattern is reflected in seasonal climatic changes, and there is an obvious relationship between the length of the day, the height of the culmination of the Sun, and the amount of light and heat received from the Sun at the place where the observations are being made. Table 1 summarizes these observations, starting from the summer.

TABLE 1 The seasonal cycle in the Northern Hemisphere

Relative lengths of daylight and darkness	Culmination point of the Sun	Name of day (date in Northern Hemisphere)	Climatic observation	Name of season that begins on this day
longest day and shortest night	highest	summer solstice (21 June)	hot	summer
day and night of equal length	average	autumnal equinox (23 September)	mild	autumn
shortest day and longest night	lowest	winter solstice (21 December)	cold	winter
day and night of equal length	average	vernal equinox (21 March)	mild	spring

As you are well aware, different parts of the Earth do not have the same season at the same time. When Britain enjoys (!) its summer, Australia has to endure its winter. In fact, the whole pattern is *reversed* for the parts of the Earth to the north and south of the Equator (the Northern and Southern Hemispheres, respectively). Furthermore, the *intensity* of seasonal changes *varies with location*. Around the Equator, seasonal changes are hardly noticeable, but in the arctic regions they are so great that even the usual day–night cycle is temporarily obliterated. There is a time (which amounts to several months at the poles) during which an arctic observer always sees the Sun moving above the horizon and never setting below it (polar day in the arctic summer) and a similar period during the arctic winter when the Sun never emerges above the horizon (polar night).

However, in spite of all these local variations, there is one common feature. For any place on Earth, the time interval between two identical observations is the same (for example, between two consecutive summer solstices, see Table 1). Thus there is some underlying regularity reflected in the *constant length of one complete seasonal cycle*, whatever its local details.

Incidentally, in order to avoid any confusion or misunderstanding that might arise from the fact that many observations are different at different parts of the Earth's surface, let us agree, from now on, that any observations described or explained will be those that can be made in Britain or in places of similar latitude within the Northern Hemisphere, unless specifically stated otherwise.

3.3 THE SUN AND THE MOON

The Sun and the Moon appear to be the two largest natural objects visible in the sky, and are the *only* two objects that appear to the unaided eye as large circular discs. All other celestial bodies visible at night appear just as small points of differing brightness.

The first interesting observation to make is that the Moon, when seen full, appears to be of about the same size as the Sun. But you know from everyday experience that the *apparent size* of an object does not tell you anything about its *real size*, unless you also know the *distance* of that object from you. A matchstick held at arm's length appears longer than the mast of a television transmitter a few miles away! At this stage, we are not concerned about the real sizes of the Sun and the Moon, or about measuring the distances to them. All we need is the observation that *the apparent sizes of the Sun and the (full) Moon each remain constant in time.** This means that whatever their distances from the Earth, these distances cannot be changing to any great extent.

The Sun is the source of nearly all the light and heat that reaches the surface of the Earth. The Moon, by contrast, is much less bright and provides no heat. It can be seen more clearly during the night, although it is frequently above the horizon during the day. The most striking feature of its appearance is that, unlike the Sun, the Moon is not always seen as a full circular disc. It undergoes a cycle consisting of **lunar phases**, as shown in Figure 7. These changes of shape correspond to the changes in the time of day or night during which the Moon can be seen above the horizon (provided that it is not obscured either by clouds or by bright sunshine). The whole cycle from one new Moon to the next is spread over approximately 28 day–night cycles.

LUNAR PHASE	TIME WHEN VISIBLE
new Moon	**during the day, near the sun**
waxing crescent	most of the day and early evening
first quarter	**afternoon and first half of the night**
waxing gibbous	evening and most of the night
full Moon	**all night**
waning gibbous	most of the night and early morning
last quarter	**second half of the night and morning**
waning crescent	before sunrise and most of the day
new Moon	**during the day, near the Sun**

FIGURE 7 Phases of the Moon and its visibility above the horizon.

* This is certainly true within the accuracy of observations made by unaided eyes or with the help of small telescopes. The fact that the Sun sometimes *seems* bigger when it is very low above the horizon has nothing to do with its distance. In part, it is due to the bending of light in thick layers of dense atmosphere, but the impression is reinforced by psychological effects.

POLARIS
CONSTELLATION
PLANET
RETROGRADE LOOP
PERIODIC PROCESS
PERIOD
THOUGHT EXPERIMENT
EXPERIMENTAL DATA (p. 20)

Another interesting characteristic of the Moon is the appearance of the darker areas seen on its surface. The shape of these features (the 'Man in the Moon') does not change appreciably with time. More important, their relative positions on the face of the Moon also appear to be constant. They temporarily disappear as one part of the Moon progressively darkens, but they reappear in the same places when that part of the Moon brightens up again. The Moon appears always to show *the same face* to the Earth.

3.4 THE STARS AND THE PLANETS

During the night, the stars appear to be moving across the sky. Closer observation reveals that in the Northern Hemisphere there is one particular star, **Polaris**, that does not seem to be moving at all. For this reason, its position in the sky has been used through the ages as a reference point for navigation.

Although some of the stars appear to be moving faster than others, there is a fascinating degree of coordination in their motions. Each individual star travels in a *circle* with respect to Polaris. Furthermore, stars are grouped at all times in the same relative configurations, known as **constellations** (*stella* is the Latin for star); two well-known examples are The Plough and Orion. Constellations appear to be remarkably stable: by comparing the present appearance of the constellations with those of the past (as recorded by ancient astronomers), we can deduce that there have been only very small changes in the shapes of some constellations over thousands of years.

By contrast with this relatively steady and highly regular pattern of the stars, there are a few very bright star-like objects that appear to move across the sky without having a fixed relationship to any of the constellations. Since they appeared to be 'stars on the loose', so to speak, the Ancient Greeks named them **planets**, meaning 'wandering stars'.

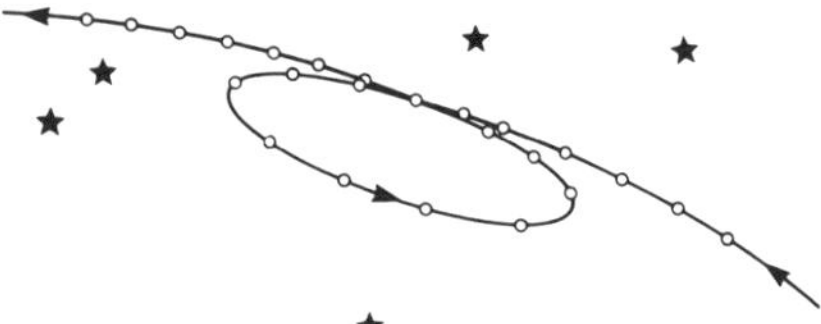

FIGURE 8 A retrograde loop in the path of a planet with respect to a chosen constellation, as seen from Earth. (Not drawn to scale.)

The most unusual feature in the behaviour of the planets is the shape of their paths with respect to the constellations. If you were to plot the relative position of a planet as seen from Earth (such as the position of Mars or Jupiter with respect to a constellation), the path could look something like Figure 8, where each successive small circle represents the relative position of the planet at the same time on successive nights. This is nothing like a segment of a circle! These loops, during which a planet appears to be turning back for a while, are accordingly called **retrograde loops**. The absence of any obvious reason for such behaviour was a weakness of the geocentric model, and this weakness led eventually to the model's demise.

3.5 THE PERIODICITY OF EVENTS

In spite of their local variations, all the observations discussed earlier in this Section indicate that there is a basic underlying regularity and an intrinsic order in the Universe. Night always follows day, spring follows winter, the first quarter of the Moon follows the new Moon, and so on.

If you choose one particular event or situation, for argument's sake the full Moon, you will observe that after a sequence of changes this particular event or situation will be repeated. Moreover, the sequence of events between the two identical situations will also be repeated. A process in which an event is repeated at regular intervals is called a **periodic process**, and the time taken to complete the sequence *once* is called the **period** of the process.

In everyday life, there are many examples of periodic processes, for example the rotation of a turntable and the swinging of a clock's pendulum. Furthermore, as you will see not just in this Unit but in the rest of the Course, periodic processes are very important throughout science. So it is worth taking a little time now to derive a better feeling for the ideas involved in the concept of periodicity. To this end, we should like you to consider an experiment. It will not involve any actual apparatus; instead, you will do it by visualizing the apparatus and by considering theoretically a situation that we shall describe. This is an example of a **thought experiment**.

Scientists are generally very fond of thought experiments, not least because they are very convenient (and cheap). However, the performing of a thought experiment certainly cannot always be regarded as an alternative to the performing of a real experiment — in the end, a theoretical idea stands or falls according to its success in accounting for the results of real experiments.

3.5.1 A THOUGHT EXPERIMENT: THE ROTATION OF A COMPACT DISC

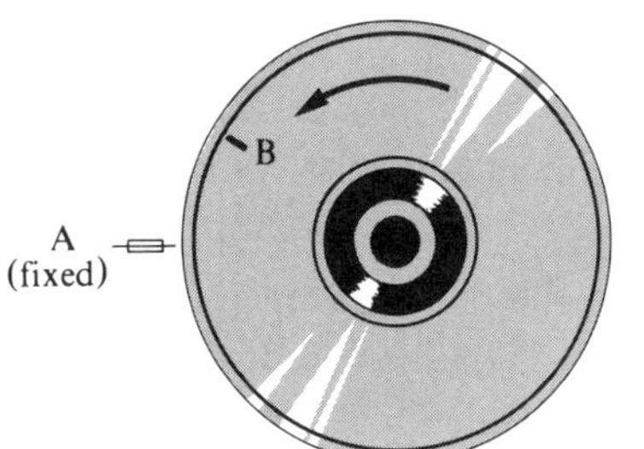

FIGURE 9 The rotation of a compact disc.

Imagine a compact disc rotating in a player (Figure 9). Think of two reference points — one just *outside* the disc on the stationary base of the machine (point A in Figure 9), and the other *on* the rotating disc (point B). In ITQ 2, we ask you to consider this imaginary apparatus.

ITQ 2 Assume that the compact disc rotates at a constant rate of 5 revolutions per second.

(a) How long does it take for the disc to complete 10 revolutions?

(b) How long does it take for the disc to complete 1 revolution?

(c) What is the period of rotation of the disc?

3.5.2 INVESTIGATING THE PERIODS OF PENDULUMS

From ITQ 2, we hope you have begun to get a feel for periodic processes. Now, we should like you to extend your understanding by doing a *real* experiment! As with all experiments in the Course it is best to read through the instructions (as far as the end of the text in the box) *before* doing the experiment.

EXPERIMENT 1 INVESTIGATING THE PERIODS OF PENDULUMS

TIME

This experiment takes about 30 minutes

NON-KIT ITEMS

(to obtain yourself)

three lengths of thin string — at least 1.2 metres, 70 centimetres and 50 centimetres long, respectively

two small heavy objects of different weights, e.g. two bunches of keys, to which you can easily tie a string

watch that can measure time in seconds, or a stop-watch

Notebook (*Note:* you will require this for *every* experiment in the Course, for recording your results.)

KIT ITEMS

None required for this experiment.

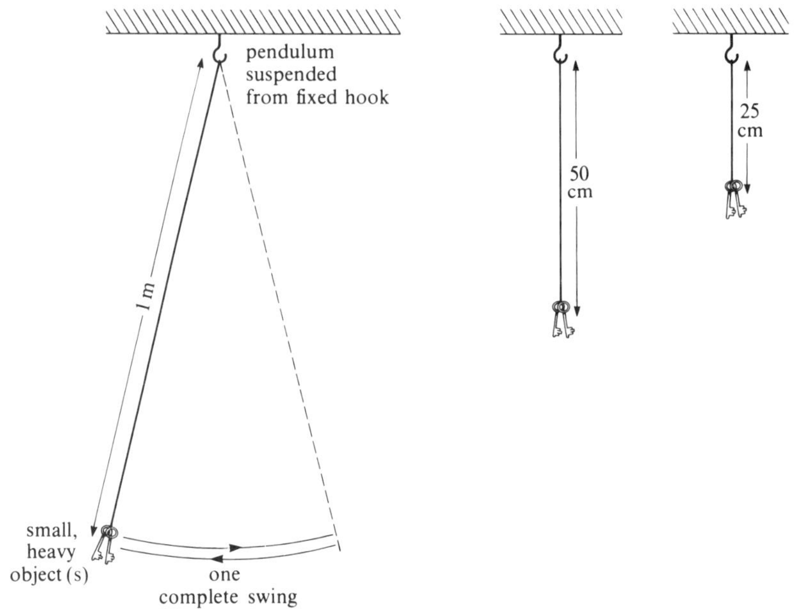

FIGURE 10 Apparatus for Experiment 1.

METHOD

Part 1

How does the period of swing vary, when you alter the length of a pendulum?

1 Refer to Figure 10. Take a piece of thin string, about 1.2 metres long, and tie the lighter of your small heavy objects to one end. (A weight at the end of a pendulum is usually called a bob.)

2 Tie the other end to a fixed point (such as a hook), taking care to ensure that the *total* length of the pendulum (from the bottom of the hook to the middle of the weight) is within a few millimetres of *1 metre.* Make sure that the pendulum can swing freely. (If you really can't find any other way of supporting the pendulum, you can, as a last resort, simply hold the free end in your hand.)

3 Copy Table 2 (below) into your Notebook, ready to enter your **experimental data**, i.e. the measurements and observations you obtain when you do the experiment. (You will need the column headed 'Period (in seconds)' later.)

TABLE 2 Results of Experiment 1

		Time for 10 swings (in seconds)			
Light or heavy bob	Length of pendulum	1st reading	2nd reading	average: (1st + 2nd)/2	Period (in seconds)
light	1 m				
light	50 cm				
light	25 cm				
heavy	1 m				

4 Hold the watch with a second-hand so that you can see the pendulum and the watch at the same time. (If you have a watch with a stop-watch facility, use this for measuring the time elapsed — and you won't need to hold the stop-watch where you can see it.)

5 Start the pendulum swinging. Measure the time it takes the pendulum to execute 10 *complete* swings, and enter this in the top row of the '1st reading' column in your Table of results. Repeat this procedure, and enter the new result in the '2nd reading' column of the top row in your Table; your two results should, of course, be identical or similar. Calculate the *average* time it takes the pendulum to execute 10 complete swings (the average time is calculated by adding the two recorded times and then dividing the result by two). Record this time in the 'average' column of the top row in your Table.

6 Remove the bob of the pendulum (i.e. the object at the end) and put the string to one side (*you will need it for Part 2, and for Experiment 2*).

You are now going to repeat the experiment with two more pendulums (with lengths 50 and 25 centimetres).

7 Make another pendulum of total length 50 cm, using the same object and the 70 cm length of string. Measure the time this pendulum takes to make 10 complete swings. Record your result in the '1st reading' column of the second row in your Table. Repeat the procedure with the same pendulum, recording the result in the '2nd reading' column. Then calculate the average time as before, recording your result.

8 Use your third piece of string (and the same object) to make a pendulum of total length 25 cm. Follow the same procedure to obtain an average time for 10 complete swings and record all your results in the third row in your Table.

9 Calculate the periods for each of the three different lengths and complete the first three rows of the Table.

Part 2

If you increase the mass at the end of your 1 metre pendulum, what will be the effect on the period?

You are now going to do the experiment with a 1 metre pendulum again, but this time use a *heavier* object as a pendulum bob (e.g. add the second bunch of keys).

10 Take this heavier object, and the 1.2 metre length of string, and repeat procedures 1–6 of Part 1; but this time record your results in the final row of your Table. (Remember to keep the length of the pendulum as close to 1 metre as possible.)

(*Again, keep your 1.2 metre length of string: you will need it later for Experiment 2.*)

Note: In every experiment in the Course, when you reach the end of the boxed text this signifies that you have finished the practical part of the experiment and can put away the apparatus. It is time to analyse your results.

ITQ 3 How does the time taken to complete 10 swings change when you shorten the length of the pendulum?

ITQ 4 What effect does shortening the pendulum's length have on the period of the pendulum's swing?

ITQ 5 When you increased the mass of the bob on your 1 metre pendulum, was the effect on the period as you expected?

3.5.3 PERIODICITY IN NATURE

Let us now return to the main story. Note that the determination of the period of a periodic process *assumes* that it is possible to measure time. However, the scientific concept of time and the techniques of time measurement are themselves firmly based on the periodicity of some natural processes. Thus, one of the first units of time was called a solar day and was defined as the interval from one culmination of the Sun to the next. Similarly, one could define a lunar month as the time between two consecutive full-Moon phases, and a solar year as the time between two consecutive summer solstices. These are perfectly adequate units for everyday use. But when it comes to the need for high accuracy they are not so good. It is rather tricky to establish *exactly* when the Sun reaches its culmination point or exactly when there is a full Moon. Also, when two solar days at different times of the year are compared very carefully, they are found not to be identical — there are small, but measurable, differences. For more accurate timekeeping, scientists nowadays look to other processes whose periodicity is more reliable. We shall look into this in much more detail in the next Unit.

SUMMARY OF SECTION 3

1 The periodic cycle of days and nights is determined by the relative motion of the Sun and the Earth. The precise details of this cycle vary, depending on where the observations are made on the Earth's surface.

2 There are longer-term, seasonal changes in the relative lengths of daylight and darkness within one solar day. These seasonal changes are related to the culmination point of the Sun and also depend on the place of observation.

3 The Sun and Moon are of roughly the same apparent size. Their distances from the Earth (not yet discussed in this Unit) do not change considerably with time (otherwise their apparent sizes would be observed to change).

4 The Moon exhibits a periodic cycle of lunar phases and has some apparently stable structural features on its surface. It appears always to show the same face to the Earth.

5 In the Northern Hemisphere, the stars appear to move in circles around Polaris.

6 Against the background of constellations, the paths of the planets (as observed from Earth) follow retrograde loops.

7 Astronomical observations show periodicity, and this enables predictions to be made even without any understanding of the system of the Universe. Periodic processes serve as a means for measuring time.

The aim of this Section was to summarize the observational evidence that can be used for the construction of a model describing the relationship of the Earth to the Sun and the Moon. In addition, you should have gained some understanding of periodic processes and have had your first taste of interpreting the results of simple observations and experiments.

4 THE EARTH, THE SUN AND THE MOON: BUILDING A MODEL

You saw in the previous Section that the Earth, Sun and Moon cannot be stationary with respect to one another. There is undoubtedly motion: but what is moving and how?

In most everyday situations, when you move (or are moved) you have a sensation of motion. But if you sit tight on a piece of rock, or anything else firmly attached to the surface of the Earth, there is no sensation that the Earth as a whole is moving. It was, therefore, only natural for most of our ancestors to believe that the Earth is stationary and that everything else is moving around it.

But is it reasonable to rely on human senses for the detection of motion? How sensitive or reliable is our sensation of motion? Consider the following imaginary situations:

1 You are travelling in a train on a perfectly straight and level track. You have just woken up from a short nap, and the first thing you see through the window is another train, just like yours, passing by. Do you think that you would be able to distinguish instantly — and without reference to other observations such as the landscape, buildings etc. — between the following possibilities:

(a) your train has stopped at a station, through which the other train is passing;

(b) your train is moving past another train that has stopped at a station;

(c) both trains are moving, in opposite directions;

(d) both trains are moving in the same direction, but one of them is travelling very much more quickly?

2 You have entered the 'cage' of a lift on the thirtieth floor of a sixty-floor skyscraper. You have not checked the light signals, did not register the initial jerk as the lift started, and the lift is now moving at a steady speed. Can you be absolutely sure at any moment whether you are travelling up or down? When will you be sure and what would you feel?

Well, the answer to these questions is, in general, 'no' — the human senses cannot always distinguish between rest and motion. The only situations in which you have a very clear sensation of motion are those in which *the state of motion is changing.* Thus, the speeding-up of the lift as it starts to ascend makes you feel pressed against the floor, and the slowing-down before it reaches the upper floor makes you feel as though you are floating upwards. The feeling is caused by the relative motion of some internal organs (the stomach in particular) with respect to the more rigid frame of your body. The sensitivity is slightly greater for up and down changes than for sideways motions, but it is generally true that a *steady* motion, in the same direction and at the same speed, is undetectable. And it is worth stressing that this is not because of any particular inadequacy of the human senses. There is no experiment and no instrument that could detect the steady motion of any system *from within that system.* This is one of the most fundamental features of the world, verified by countless experiments.

But what about motion in a circle, such as on a rotating turntable or a roundabout? Well, the problem here is that this is not a steady motion. Even if the *speed* of rotation is constant, the *direction* of motion is continuously changing, and you can feel this motion when you sit on a rotating roundabout. But even here, human senses are not only fallible but also very adaptable. It is well known that after a sufficiently long exposure to some roughly constant influence — smell, noise, pressure, etc. — people can become so used to it that they actually become unaware of it. It is not very difficult to imagine that if someone spent most of their life in the seat of a smoothly and steadily rotating roundabout, without any reference to the

SPIN OF AN OBJECT

AXIS OF ROTATION OF AN OBJECT

AXIS OF SPIN OF AN OBJECT

ORBITAL CIRCULAR MOTION

world outside, they would not have any sensation of rotation. Although such motion is always detectable by suitably designed experiments, it is not necessarily always registered by human senses and the human brain.

Thus, when it comes to answering the question of how the Earth relates to the rest of the Universe, there is *no justification for assuming that the Earth itself is not moving* — it may be or it may not.

4.1 THE EARTH AND THE SUN

In order to understand the place of the Earth in the Universe, it is necessary first to understand its relationship to the Sun. The point has already been made in Section 3 that this relationship cannot be observed directly, but it can be represented by a suitable model.

It is reasonable, in view of the arguments advanced in Section 2, to start by representing the two bodies by two spheres. The sphere that represents the Sun can be assumed to radiate heat and light, and moreover we can assume that equal amounts of heat are radiated in all directions (and the same for light). Further developments of the model will consist of choosing appropriate relative motions of the two spheres that would *reproduce the observations* described in the previous Section. (Note that quantitative details of the model, such as the actual sizes and separation of the two bodies, will be deferred until Unit 2.)

What kinds of possible relative motion should you consider for the two spheres? If you recall the observation that both the day–night cycle and the cycle of the seasons are (in spite of all their local variations) essentially periodic processes, it is obvious that the relative motion of the two bodies must be *periodic*. That is, whichever of the two bodies is moving and however it moves, there can be only a limited sequence of possible different configurations of the two bodies and this sequence of configurations is regularly repeated. Furthermore, the observation that the apparent size of the Sun, as seen from the Earth, is always the same, can only mean that the two spheres must always be at the *same separation*.

□ What type of motion satisfies these conditions of periodicity and constant distance of separation?

■ It has to be motion in a circle. You might have thought that the motion of a pendulum bob also satisfies these conditions. But this can be ruled out because a pendulum bob periodically reverses its direction of motion, whereas the Sun always moves in the same direction.

Thus, by using two observations, you already have one necessary feature of any model for the Sun–Earth relationship:

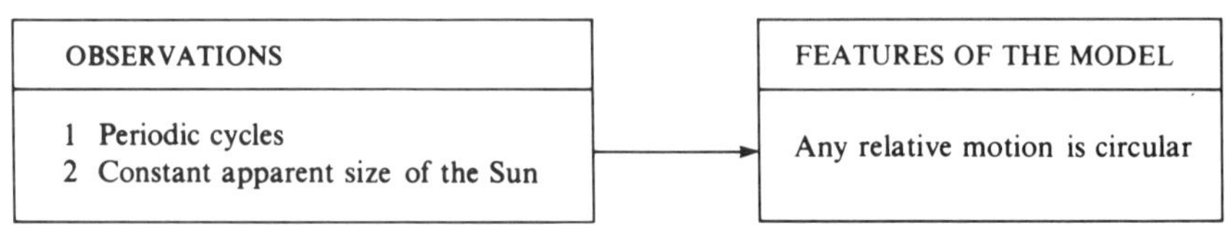

4.1.1 TWO KINDS OF CIRCULAR MOTION

At this point, we must be careful to differentiate between two kinds of circular motion — spinning, and orbital circular motion.

When a body rotates about a (normally imaginary) fixed line that *passes through* the body, the body is said to be **spinning**, and the fixed line about which it is turning is called the **axis of rotation** (or sometimes the **axis of spin**). Examples of objects that execute spinning motion are the drilling bits in power tools, a child's spinning top, and pirouetting ice-skaters. Perhaps you can think of more.

On the other hand, when a body moves in a circular path about a point, the body is said to be executing **orbital circular motion**. Note, however, that it is conventional to use the term orbital motion also when the body follows a path that is not circular, but which nevertheless has a fixed, closed shape. In this way, the moving body will travel repeatedly, in the same way, during each complete revolution. The most common non-circular orbital path is the ellipse, an oval-shaped path which will be considered in more detail in Unit 2.

Common examples of orbital motion are greyhounds running around a closed race-track, seats on a merry-go-round, and a satellite orbiting the Earth.

Returning to our model of the two spheres representing the Sun and the Earth, there are many possible combinations of different motions. Either of the two bodies can orbit the other, and each of them may or may not spin. This variety of options can be reduced by taking into account the *appearance of the Sun.*

Does the Sun show any periodic changes of its shape or of its brightness from which one could conclude that it spins? No such changes have been noticed in everyday observations. Large telescopes reveal some very small spots of lower brightness on the Sun's surface (sunspots). The motion of these spots indicates that the Sun could be spinning with a period of about a month but, in spite of this, the overall amount of light and heat emitted from the Sun's surface remains the same in all directions and at all times (within a typical human lifespan, at least). It is therefore reasonable to ignore any spinning of the Sun, because it does not make any difference to Earth-bound observations.

This leaves only four different options for the motions of the Sun and the Earth, as shown in Figure 11.

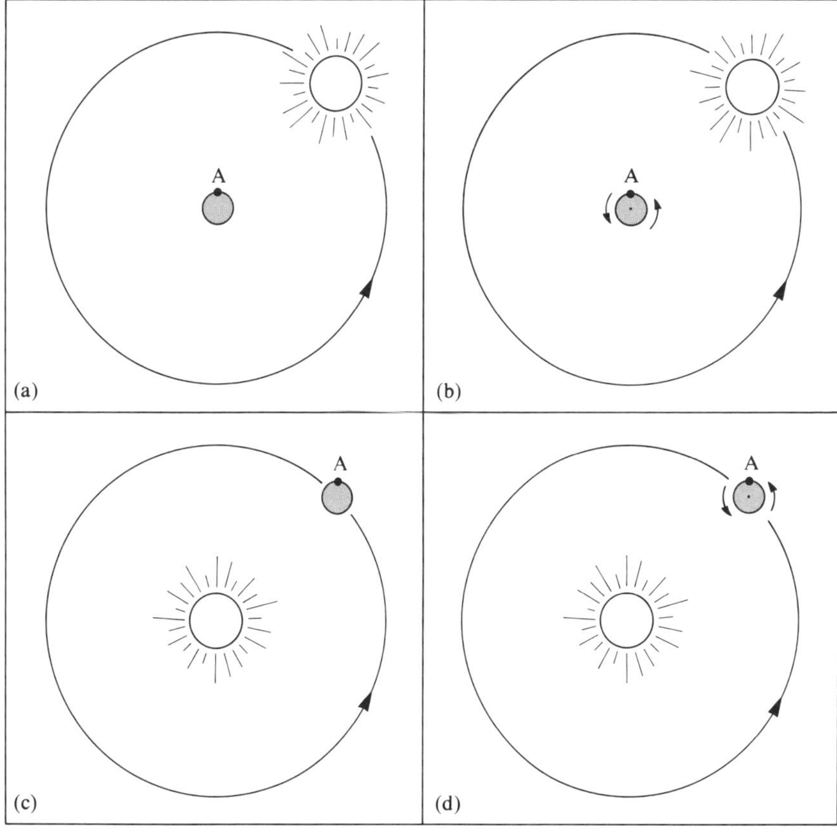

FIGURE 11 Possible options for the model of the Sun–Earth relationship. The dot labelled A represents an observer on Earth. The dot at the centre of the Earth, in (b) and (d), indicates the Earth's axis of rotation.

FOUCAULT PENDULUM EXPERIMENT

The models in Figure 11 may be described as follows:

(a) The Earth is stationary and does not spin; the Sun orbits around the Earth.

(b) The Earth is stationary and spins; the Sun orbits around the Earth.

(c) The Sun is stationary; the Earth orbits around the Sun and does not spin.

(d) The Sun is stationary; the Earth orbits around the Sun and also spins.

(Note that stationary in this context does not mean motionless. It means a fixed position in space; spin *is* allowed.)

4.1.2 WHAT CAUSES THE DAY–NIGHT CYCLE?

Leaving aside everything else for the moment, option (a) could account for the day–night cycle, provided that the Sun could complete one full orbital cycle in one solar day. This would lead to very large, perhaps even impossibly large, speeds for the Sun, if its distance from the Earth is very large. Although this is a potential problem, it would not on its own justify the rejection of this option. A far more serious objection to (a) is that, having used the orbit of the Sun to explain the day–night cycle, there is no other motion left to account for the periodic seasonal cycle.

A very similar argument can be made against option (c). Once again, a day–night cycle could be accounted for by the Earth's completing one circle in one solar day, but there is no other motion left to explain any other periodic observations (the seasons).

Options (b) and (d) have the obvious advantage that two different motions are available and therefore more than one periodic observation can be explained.

But perhaps there might be some other explanation for the long-term seasonal cycle, not necessarily dependent on the spin or orbital motion? It does not seem likely, but in order to reject options (a) and (c) convincingly we require some direct evidence that they are wrong.

□ Identify the *assumption* about the Earth that is common to options (a) and (c). What kind of an experiment would eliminate both of these options?

■ The common assumption is that the Earth does not spin. Therefore, any experiment that shows the Earth does spin, conclusively eliminates both options (a) and (c).

So, is there any way to check whether the Earth spins? We'd like you to begin to consider this question by trying another pendulum experiment.

EXPERIMENT 2 INVESTIGATING THE DIRECTION OF THE SWING OF A PENDULUM

TIME

About 10 minutes

NON-KIT ITEMS

thin string, about 1.2 m long (you can use the longest piece of string from Experiment 1)

a small, heavy object, e.g. one of those from Experiment 1

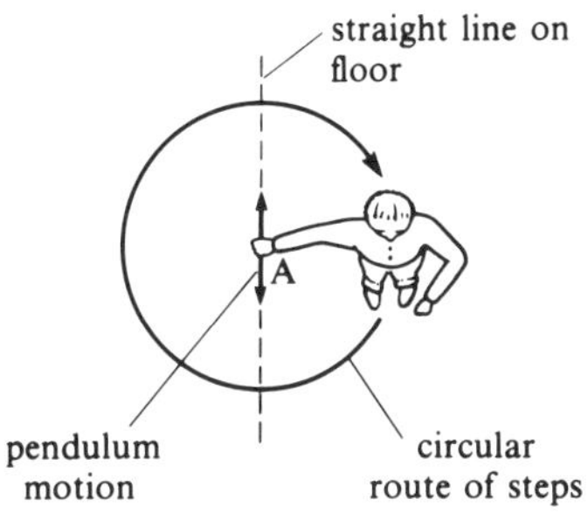

FIGURE 12 Experiment 2.

METHOD

1 Using the string and the heavy object, make a pendulum of length 1 metre.

2 Find a fixed line on your floor, utilizing the pattern of your carpet, a row of tiles or something similar. Suspend the pendulum from your fingers and start it swinging along this fixed line. Keep your arm outstretched and imagine that your fingers holding the pendulum are at the centre of a circle.

3 Gradually move around in a complete circle, keeping the point of suspension of the pendulum fixed (Figure 12). Observe the motion of the pendulum with respect to the reference line.

ITQ 6 As you moved round in a circle, did you observe any change (with respect to the fixed line on the floor) in the pendulum's direction of swing? What do you conclude?

Now imagine what would happen if, instead of walking in a circle around the swinging pendulum, you were standing on a rotating platform and held the end of the pendulum string above the centre of the platform. As far as the pendulum is concerned, the situation is exactly the same as before: the swinging object is freely suspended* from a rotating arm. So, in the rotating-platform case, the pendulum will always swing in the same plane, just as it did in your experiment.

□ Does this suggest to you a possible experimental test that would show whether the Earth is spinning?

■ All you would have to do would be to use a very heavy bob on a very long pendulum, suspended with a minimum of friction, so that it would go on swinging for several hours, once started. You would draw a reference line on the surface of the Earth, indicating the initial plane of swing, and then you would just sit and wait. If the Earth is not spinning, the pendulum would never deviate from this line. But what if the Earth, carrying the line, spins underneath the pendulum? After a sufficiently long time, there would be an obvious difference between the reference line and the plane of swing.

This is the basis of the **Foucault pendulum experiment**, which is demonstrated in the TV programme associated with this Unit, 'The planet Earth — a scientific model'. As you will see in the TV programme, this experiment

* By 'freely suspended' we mean that the motion of the bob is not restricted by any *rigid* coupling to the point of suspension. A long, thin string has very little restraining effect even if it is held fixed or tied up to a hook. However, if the pendulum had been made using a rigid rod instead of a piece of string, then some special suspension would have had to be designed in order to enable the rod to swing without friction in any direction.

enables us to reject options (a) and (c) because it shows that the Earth does indeed spin.

The Foucault pendulum not only allows us to reject options (a) and (c), it also helps us to clarify one important detail in options (b) and (d). In these options there are two periodic motions — the spin of the Earth, and the orbital motion of either the Sun (b) or the Earth (d); we need to connect these with two periodic observations — the day–night cycle, and the seasonal cycle. It may seem *intuitively* more reasonable to connect the spin of the Earth with the shorter of the two cycles, but it is the outcome of the Foucault pendulum experiment that provides a clear answer. Even without accurate measurements, it is obvious from the rate at which the plane of swing deviates from the reference line on the Earth's surface that the spin of the Earth accounts for the day–night cycle rather than for the seasonal cycle. In fact, a Foucault pendulum suspended exactly above the North Pole would show that *one turn of spin of the Earth takes 24 hours.* In Britain, it takes about 30 hours for the reference line to complete one full turn with reference to the plane of swing of a Foucault pendulum. The difference is due to Britain's distance from the pole. The explanation of this difference is based on the fact that the line between the point of suspension and the bob of a stationary pendulum always points towards the centre of the Earth — the reason for this will become clear in Unit 3. For a pendulum suspended over the pole, the line of suspension *coincides* with the axis of rotation of the Earth. This is why the rate of deviation between the initial reference line and the plane of swing of a Foucault pendulum at the pole measures *directly* the rate at which the Earth is turning about its axis of rotation. Above the Equator, on the other hand, the line of suspension of the pendulum is *perpendicular* to the Earth's axis of rotation and the Foucault pendulum would not show any deviation from the initial reference line. In Britain, the rate of deviation is slower than at the pole, because the line of suspension is at an angle of about 35 degrees (often written as 35°) to the Earth's axis of rotation.

Before taking the next step in our model-building exercise, let us sum up our progress so far.

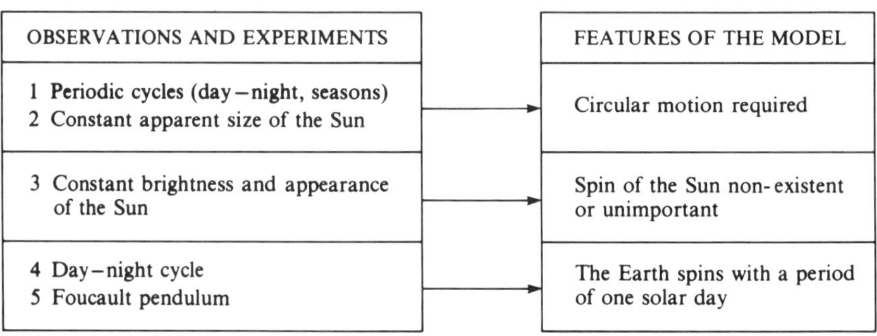

4.1.3 WHAT COULD EXPLAIN THE SEASONAL CYCLE AND ITS LOCAL VARIATIONS?

In Section 4.1.2, the spin of the Earth was firmly linked with the day–night cycle. Therefore if option (b) or (d) is to explain the relative motion of the Earth and the Sun, it follows that the second periodic observation — the cycle of the seasons — could be connected only with the orbital motion. Either the Sun orbits around the Earth — option (b) — or the Earth orbits around the Sun — option (d) — and in either case the period of the orbital motion has to be exactly the same as that observed for a complete cycle of the four seasons — one solar year.

But there is a snag. Can you see what it is?

□ What is the shape of the orbit (according to all the evidence so far)? Does this shape tie up with the sequence of seasonal changes?

■ The orbit is circular (evidence for this is provided by the constant apparent size of the Sun). But if the two bodies are always at a constant separation, how can the amount of light and heat received from the Sun vary so much from one season to another?

Your everyday experience tells you that the nearer you stand to a fire, the more light and heat reaches you. Thus it seems reasonable to look critically at the conclusion reached earlier about the circular orbits. Perhaps there are some changes in the distance between the Sun and Earth after all? Perhaps it is just possible that these changes are large enough to explain the variations in the amount of light and heat that we receive on Earth at different times, and yet small enough not to make a noticeable change in the apparent size of the Sun.

Figure 13 illustrates two possible modifications that could be introduced into the relative motion of the two bodies. Note that in both cases the axis of rotation of the Earth is assumed to be perpendicular to the plane of the orbit (which is in the plane of the paper).

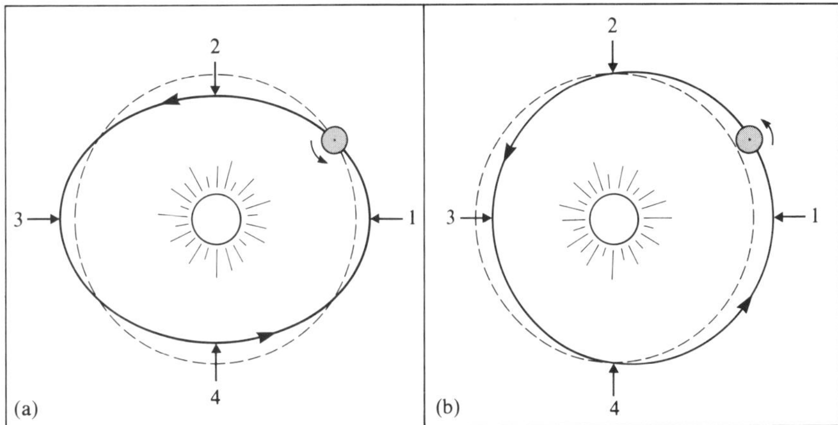

FIGURE 13 An attempt to explain the seasons by (a) altering the shape of the Earth's orbit, or (b) shifting the orbit so that the Sun is no longer at its centre. In both cases, the dashed line shows the unmodified orbit.

□ Would it make any difference if the Earth and the Sun changed places in Figure 13, so that the Earth was at the centre and the Sun orbited round it?

■ No, it would make no difference, because after the bodies had changed places, their separation would remain the same.

ITQ 7 Follow the progress of the Earth through the arrowed positions in Figures 13a and 13b, and identify the season you would expect the Earth to experience at each of them.

So far so good. But what about the fact that different parts of the Earth experience *different* seasons *at the same time?* There is no way that this can be explained by the modifications to the orbit shown in Figures 13a and 13b. Indeed, we seem to have reached an impasse. On the one hand, some evidence suggests that the orbit ought to be circular, or at least very nearly so, whereas other evidence seems to indicate that some parts of the Earth are nearer to the Sun than others. Surely, this is impossible?

Well, not quite. There is still one important aspect in the relative configuration of the two bodies that has not been taken into account properly. You probably know what it is, or can easily guess. It is the relative orientation of the Earth's axis of rotation to the plane of the Earth's orbit.

In all previous discussions, and particularly in Figures 11 and 13, the Earth's axis of rotation was *assumed to be perpendicular to the plane of the orbit* (of the Earth or of the Sun). Perhaps this assumption was wrong? What if the axis of rotation is actually oriented so that it lies *within the plane* of the orbit? Would that make any difference?

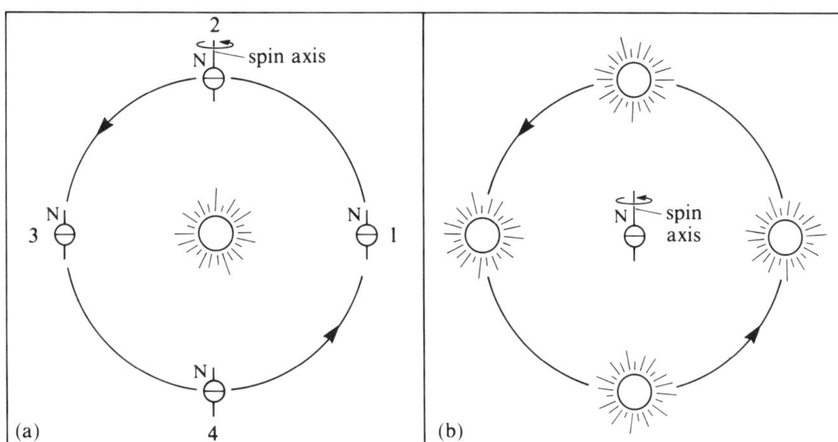

FIGURE 14 Can the axis of rotation lie in the plane of the orbit? (a) The Earth orbiting the Sun; (b) the Sun orbiting the Earth.

Well, have a look at Figure 14 and answer ITQs 8 and 9 by considering what an observer in Britain (i.e. in the northern Earth hemisphere in the Figure) would experience, as the Earth spins about an axis lying in the plane of the orbit (i.e. in the plane of the paper).

ITQ 8 Concentrate first on the situation depicted in Figure 14a. Summarize briefly what the observer in Britain would experience as the Earth moves through the numbered positions indicated on the orbit. Consider both the day–night cycle and the seasonal variations.

ITQ 9 Compare Figure 14a with 14b, and identify in Figure 14b the situations in which the observations for an observer in Britain would be identical with those in the numbered positions in Figure 14a. Is there any observation in Figure 14a that could not be made by an observer in the Northern Hemisphere in Figure 14b?

Incidentally, you may well be wondering why the axis of rotation as shown in Figure 14 has *the same orientation in space* all the time, thus assuming different orientations with respect to the line connecting the two bodies. We do not want to go into detail about this, but it is generally true that just as the plane of swing of a freely suspended pendulum is fixed, so the axis of rotation of a freely spinning object is also fixed. If you have ever tried to spin a bicycle wheel on its axle, holding the axle in your hands, and then attempted to tilt the axis, you will have experienced the resistance of the wheel to this change. The more massive the body, and the faster it spins, the more difficult it becomes to change the direction of its axis of rotation. Hence a body spinning freely, that is without any rigid coupling to another body, will keep its axis of rotation fixed forever.

After that digression, let us return to our main problem: how can we account for the seasonal cycle and for the fact that it is different at different places on the Earth?

You have just seen that adjustments of orbits do not work (Figure 13). Changing the axis of rotation from a direction perpendicular to the plane of the orbit to a direction that lies within the plane does not work either (Figure 14). However, examination of some of the situations in Figure 14 should have reminded you of the observational evidence that there are places on the Earth where, for a certain part of the year, the Sun never rises and never sets. There are only two such areas; they are situated in opposite parts of the Earth, and their experience of polar days and polar nights is complementary. Since the spin of the Earth as a whole does not change (as witnessed by the regular day–night cycle at all other parts of the surface), the only possible explanation of this effect is that the axis of rotation is not perpendicular to the orbit, and nor does it lie within the plane of the orbit. It must be *tilted* at an angle somewhere nearer 90° than 0° to the plane of

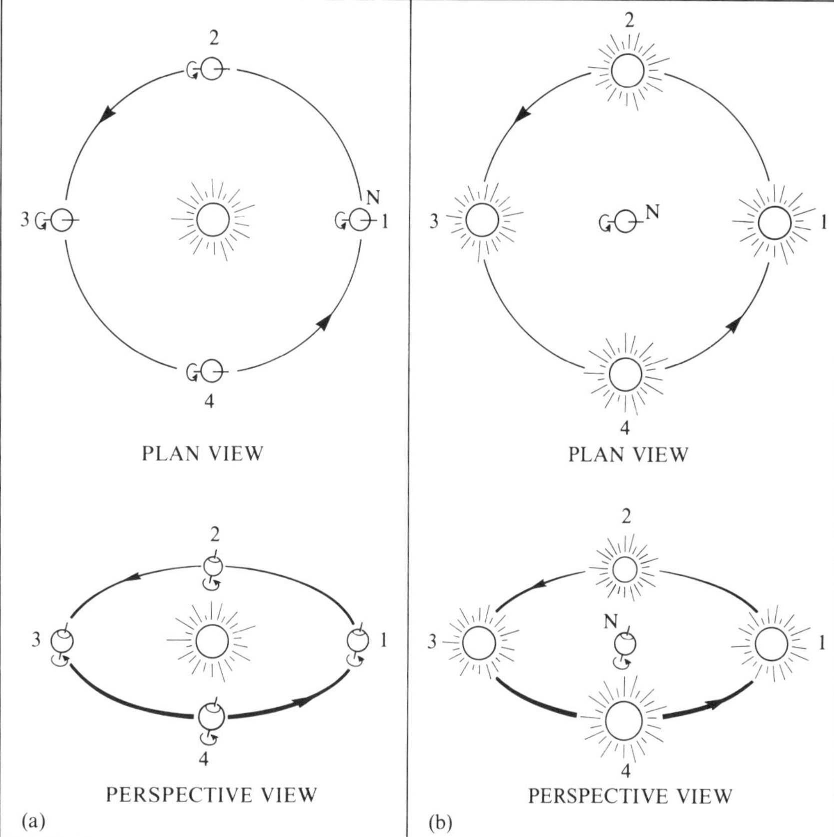

FIGURE 15 If the Earth's axis of rotation is tilted slightly away from the perpendicular to the orbit, this can account for the seasonal cycle and its local variations. (a) The Earth orbiting the Sun; (b) the Sun orbiting the Earth. N indicates the Earth's Northern Hemisphere. Note that the lower half of the Figure gives *perspective* views, in which circular orbits appear elliptical.

the orbit (at 90° there would be no polar nights anywhere, see Figure 13, and at 0° they would be experienced over the whole hemisphere at the same time during the year, see Figure 14).

Later, from an independent piece of evidence, you will be able to confirm this conclusion and give a better quantitative estimate for the angle of the tilt. In the meantime, convince yourself, by studying Figure 15, that such a tilted axis of rotation can indeed account for the existence of seasons as well as for their local variations. (If you can't manage to picture it from Figure 15 — we don't all possess a good visual imagination — it might help if you push a knitting needle through the centre of an apple (to represent the Earth) and move the apple in a horizontal circle round an orange (representing the Sun). Keep the axis of tilt — the direction of the needle — constant and give the apple a spin on the needle at the various positions, to try and find which parts of the apple would have a 'view' of the orange. Even better, if you happen to have a globe mounted on a tilted axis, use this instead of the apple and knitting needle.)

SAQ 2 Deduce from Figure 15a which season would be experienced by an observer in the Northern Hemisphere (N), and which season would be experienced by an observer in the Southern Hemisphere, as the relative position of the Earth and the Sun changes from 1 to 4. Hence complete Table 3.

TABLE 3 For use with SAQ 2

Location of observer	Configuration			
	1	2	3	4
Northern Hemisphere	winter			
Southern Hemisphere		autumn		

HELIOCENTRIC MODEL

SAQ 3 Compare Figure 15b and Figure 15a and decide which positions of the Sun in Figure 15b would give rise to the same seasons as experienced in positions 1 to 4 in Figure 15a.

SAQ 4 Incorporate the conclusions of Section 4.1.3 into boxes 6 and 7 of the summary below of the model we are developing for the Earth–Sun relationship.

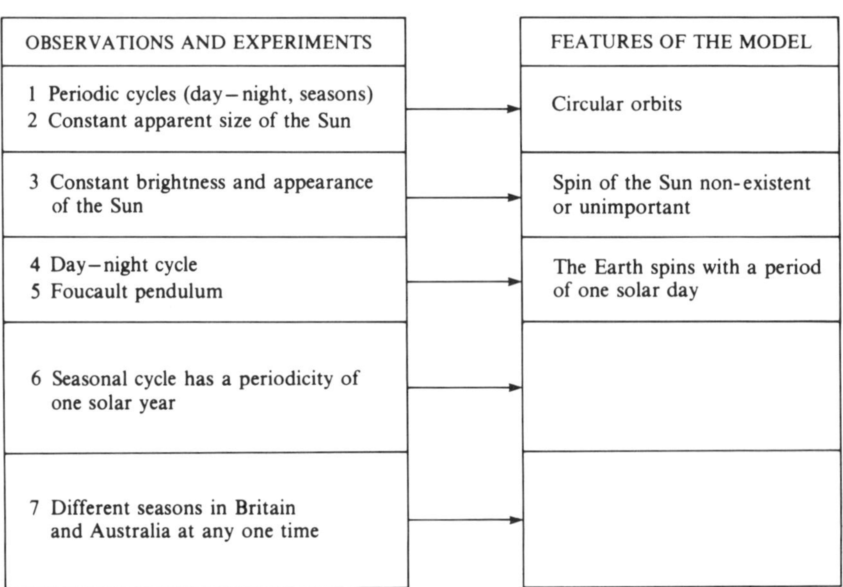

4.1.4 WHICH OF THE TWO BODIES IS ORBITING — THE SUN OR THE EARTH?

So far, we have not resolved the question of which body — the Sun or the Earth — is in orbit around the other; options (b) and (d) (Figure 11) have consistently given the same results. (You have tested this, for example, on Figures 14 and 15.) Does this mean that for Earth-bound observers there is nothing to choose between the two models, or is there any observation that indicates that one of the models is superior to the other?

Well yes, there is. You may remember from Section 3 that if the position of a planet (for example, Mars) is plotted night by night with respect to a chosen constellation, the plotted path of the planet is not uniformly in one direction. Occasionally it appears that the planet reverses its direction for a while, before returning to its initial direction (Figure 8). This behaviour would be very odd indeed, if we were to believe that the Earth is stationary and that the Sun as well as all the planets travel around it. There is no obvious reason why planets should have such complicated orbits; for example, there is no other large body anywhere near them that could temporarily deflect them from their original paths.

Could this strange behaviour be explained in a simple way if we were to use a model in which the Sun is stationary and is orbited by the planets (including the Earth)? This model is normally called the **heliocentric model** (*helios* is the Greek for Sun); but it is also referred to as the Copernican model after the Polish astronomer Nicolas Copernicus (1473–1543), who worked out in detail the motions of all the planets around the Sun.

The heliocentric model does indeed offer a very simple explanation for the retrograde motion of the planets. Figure 16a shows a heliocentric orbit of the Earth and a part of the orbit of another planet, say Mars. Numbers 1–9 correspond to simultaneous positions of the two planets. (To save space, the motion of Mars is shown 'slowed down'. In reality, Mars would cover about one-half of its orbit during one orbital period of the Earth. This does not change the principle of the explanation — you could think of the second orbit as that of a planet much farther away from the Sun than Mars,

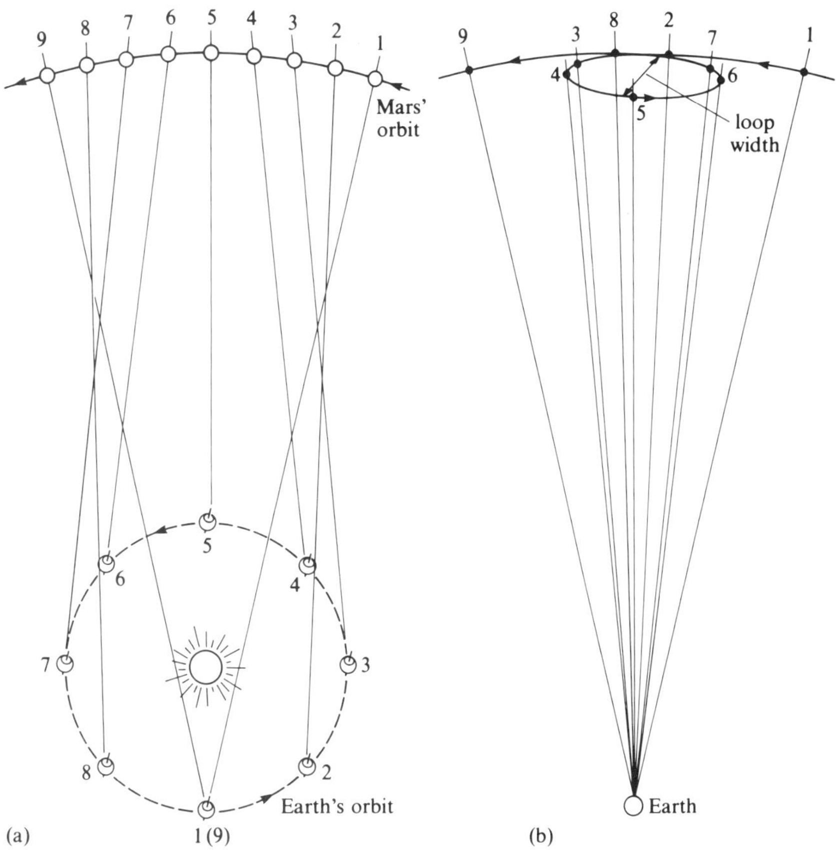

FIGURE 16 The orbit of the planet Mars.

(a) How we should see the motion of the two planets Mars and Earth with respect to the Sun, if we could look down on them from some distant point in space. Arcs 1–2, 2–3, 3–4, etc., in both orbits, represent the motion during the same fixed time intervals.

(b) The path of Mars as seen from one point on Earth. Earth-bound observers perceive that they are stationary, and therefore all the observations have been represented as if they had been made from one point on a stationary Earth, but with the directions unchanged from those in (a).

if you prefer.) The lines 1–1, 2–2, 3–3, etc. in Figure 16a define the *directions* in which the other planet is seen from the Earth, against the distant background of the stars.

When all these directions are transferred to a single point (Figure 16b), you can see that to Earth-bound observers (who believe themselves stationary) the other planet moves from right to left at first ($1 \rightarrow 2 \rightarrow 3 \rightarrow 4$), but then *reverses* its direction of motion ($4 \rightarrow 5 \rightarrow 6$), before resuming the original direction again.

Thus the heliocentric model offers a simple and natural explanation of retrograde loops: they result from a *combination* of two circular orbital motions. There is no need to look for any 'unknown' influence, as would be required in a geocentric model.

The loop width, indicated by an arrow in Figure 16b, arises from the difference between the orbital planes of the Earth and the other planet. For the sake of clarity, the loop width has been exaggerated. If all the planets orbited in exactly the same plane, the retrograde motion would not be a loop: it would appear simply as a back-tracking along the planet's previous path. The phenomenon of retrograde motion is illustrated further in the TV programme associated with this Unit.

In conclusion, we have found a good reason to prefer the heliocentric (Sun-centred) model to the geocentric (Earth-centred) model: the motion of the planets is very much simpler if the Sun, rather than the Earth, is taken to be the centre of the whole planetary system.

ECLIPSE

4.2 THE MOON AND ITS RELATIONSHIP TO THE EARTH AND THE SUN

The most striking observation about the Moon is the periodic cycle of phases described in Section 3.3. The continuously changing shape and size of the bright part rules out any possibility that the light from the Moon originates in the Moon itself. It is very difficult to imagine any physical process on the surface of the Moon, that could switch different parts on and off with such unfailing regularity and with such a sharp boundary between dark and light.

On the other hand, the Moon's low overall brightness, the negligible rate at which it emits heat (compared with the corresponding rate for the Sun), and the shape of its bright crescents can be readily understood by assuming that the Moon is a spherical body, *illuminated by the Sun and reflecting some of its light*. But how is the Moon placed with respect to the Sun and the Earth, and how does it move? What features should a three-sphere model have in order to account for all the observations?

4.2.1 WHAT PATH DOES THE MOON FOLLOW?

According to our model, the Earth moves in a circular orbit around the Sun, completing one circle in one year. As it goes around the Sun, it also spins about its axis of rotation (tilted with respect to the plane of the orbit), completing one turn in one solar day.

You also know from Section 3.3 that the apparent size of the Moon does not change. Even if you see only a narrow bright crescent, if you were to complete the circle that corresponds to the crescent, the size of the circle would be the same as that of the full Moon. Thus the distance between the Earth and the Moon must be *constant*. And there are only two ways in which this could be true. Either the Moon moves alongside the Earth in a concentric circle around the Sun, or else it moves in a circular orbit *around* the Earth and at the same time accompanies the Earth in its orbit around the Sun.

The first possibility is easily discounted. Can you see why? Look at Figure 17. In this configuration, there would be no lunar phases. The Moon would look the same at all times to an observer at any one place on the Earth, although it would look different to observers at different places. You might like to try to imagine how you would see the Moon if you were sitting somewhere in the Northern Hemisphere of the Earth and if the Moon were moving alongside the Earth. Consider six different possibilities for the position of the Moon: inside or outside the Earth's own orbit, and, for each of these, ahead, behind or exactly on the Sun–Earth line (Figure 17).

So, in order to account for the lunar phases (Figure 7), the Moon must be moving *around* the Earth. And since the whole cycle of the phases has a period of about 28 solar days, this is likely to be the time it takes the Moon to complete one circle around the Earth.

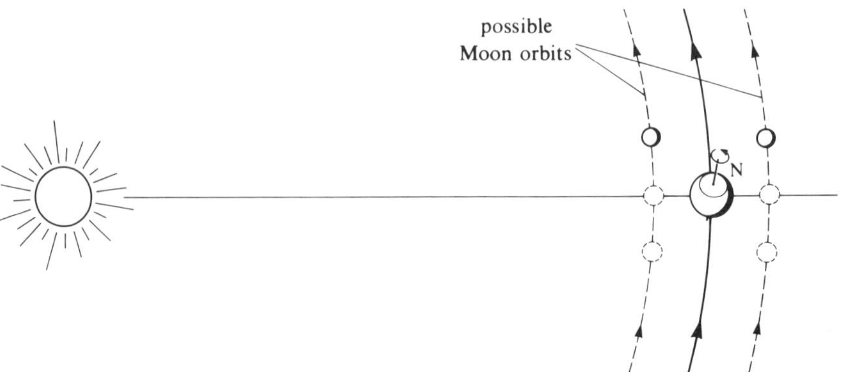

FIGURE 17 Can the Moon travel *alongside* the Earth?

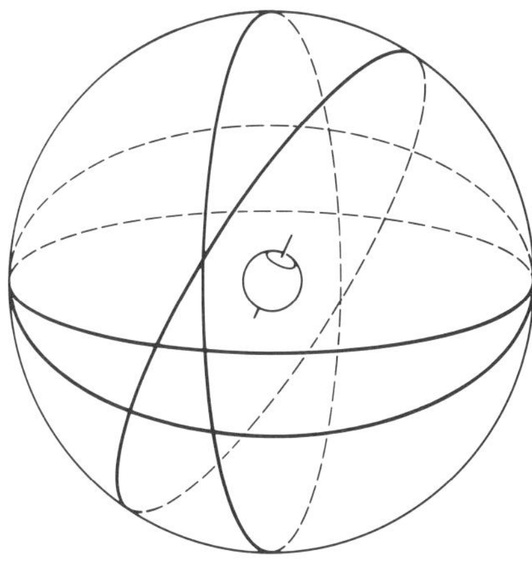

FIGURE 18 A sphere is defined by all possible circular orbits with a common centre and equal radius. (Note that the 'dashed' parts of the orbits define the 'invisible' half of the sphere.)

4.2.2 WHAT IS THE PLANE OF THE LUNAR ORBIT?

Having established that the Moon cannot simply travel parallel to the Earth but must move around it in a circular orbit, it is pertinent to ask: How is the lunar orbit related, in space, to the orbit of the Earth? This is not a trivial question, as you can see from Figure 18: there is an infinitely large number of different circular orbits with the same centre point and of the same diameter, all lying on one spherical surface surrounding the Earth. Since we cannot tell which is the most likely orbit, we may as well start by looking at two distinct options and comparing how they could account for our real observations.

Let us first assume that the plane of the lunar orbit is *identical* with the plane of the Earth's orbit. Imagine yourself to be somewhere in the Northern Hemisphere of the Earth (Figure 19) and consider how and when you would see the Moon as it travels around. (Remember, both the Earth's orbit and the lunar orbit lie in the plane of the paper.)

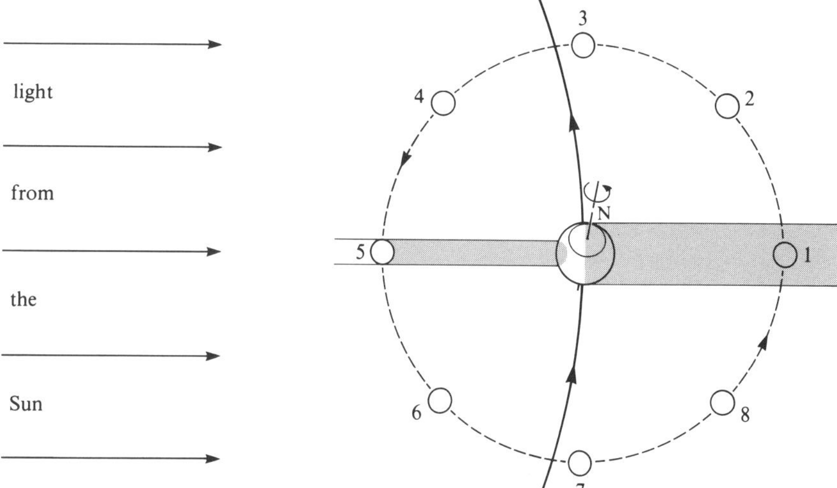

FIGURE 19 What happens if the orbit of the Moon lies in the same plane as the plane of the Earth's orbit?

When you are trying to imagine how you would see the Moon during each day (one spin of the Earth), you should note that the orbit of the Moon in Figure 19 (as well as in the later Figures) *is not drawn to scale*. The distance of the Moon from the Earth is about 30 times the diameter of the Earth — to draw this relationship to scale would be somewhat wasteful of paper! But because of this large distance, the *appearance* of the Moon will not change appreciably between, say, the evening and morning, although the observer's position has changed as the Earth spins. If you are finding it hard to visualize the likely appearance of the Moon, you will receive considerable help from the TV programme associated with this Unit.

Now try the following three ITQs, in which you will use the idea of **eclipses** — the total or partial obscuring of one celestial body by another. These ITQs are quite demanding, so don't worry if you find them difficult!

(a) **new Moon**
(b) waxing crescent
(c) **first quarter**
(d) waxing gibbous
(e) **full Moon**
(f) waning gibbous
(g) **last quarter**
(h) waning crescent
(i) **new Moon**

FIGURE 20 Phases of the Moon.

ITQ 10 Complete each of the statements (i)–(iv) below with one of the Moon's (numbered) positions shown in Figure 19. You will probably find it helpful to refer to Figure 20, which reminds you of the names of the Moon's phases.

(i) The Moon is in the waxing crescent phase (b) when in position ....

(ii) The Moon is eclipsed by the Earth when in position ....

(iii) The Moon is in the last quarter (g) when in position ....

(iv) The Moon is in the waning gibbous phase (f) when in position ....

ITQ 11 Describe the lunar phases that would be observed in the remainder of the positions shown in Figure 19.

ITQ 12 Identify by numbers from Figure 19 the situations in which the Moon is above the horizon in the Northern Hemisphere at the following times:

(i) the afternoon and first half of the night;

(ii) most of the night and early morning;

(iii) all day;

(iv) the evening and most of the night.

It is very encouraging to see that this choice of the lunar orbit (in the same plane as the orbit of the Earth) seems to reproduce the cycle of phases quite well. The only problem is that, from Figure 19, you would expect a full eclipse of the Moon *every month* (position 1). And similarly you should predict regular monthly eclipses of the Sun, when it is covered by the disc of the Moon (position 5). In practice, however, eclipses are much less frequent. So the choice could not have been quite right after all.

Well, would the other distinct option, namely that of a lunar orbit *perpendicular* to the plane of the Earth's orbit, do any better? At first sight, this is a much more complicated situation than before. Can you see why? There is only *one* way in which the lunar orbit can lie in the plane of the Earth's orbit. But there are any number of different planes for the lunar orbit that are perpendicular to the plane of the Earth's orbit. If you cannot see this immediately, imagine that the Earth's orbit lies in the plane of your work surface and the lunar orbit in the plane of the front cover of this Unit. How many ways are there for you to make the front cover perpendicular to the top of the work surface? Clearly, if you start from one such perpendicular position, you can rotate the book through a full circle and it will remain perpendicular to the top of the work surface all the way round.

In case you are worried that you are going to be asked to go through an endless number of plane orientations, don't despair. All of them can be eliminated in a stroke!

Look at Figure 21. This shows, in a perspective drawing, the plane of the Earth's orbit and an arbitrarily chosen, perpendicular lunar orbit. If you were an observer in the Northern Hemisphere, would you see the Moon, if only for a short while, during *every* spin of the Earth (each solar day)?

Clearly not, because for one-half of its way around the Earth (that is for some 14 days or so) the Moon would remain below the level of your horizon (between positions 5–6–7–8–1, i.e. the part of the orbit shown with a dashed line to indicate it is below the plane of the Earth's orbit and therefore 'out of sight'). This would be true, whichever way you turned the plane of the lunar orbit, as long as it remained perpendicular to the plane of the Earth's orbit. And this is in contradiction with our experience, for the Moon is above the horizon, at some time or other, during *every solar day*. Thus, whatever the orientation of the lunar orbit in space, it cannot be perpendicular to the plane in which the Earth travels around the Sun. In fact, the observation that the Moon is never below the horizon for more than part of a solar day at a time, clearly supports the conclusion that the plane of the lunar orbit cannot be *very* different from the plane of the Earth's orbit, although it cannot be *quite* identical with it, as you saw before.

Let us check whether a *slightly inclined* lunar orbit, as in Figure 22, can cure the problem of having too many eclipses. Note that this Figure is a *perspective* drawing, with circular orbits shown as ellipses; this is necessary to convey the impression of their relative inclination. Note also that although one-half of the lunar orbit (the part shown with dashed lines) is now below the plane of the Earth's orbit, the inclination is sufficiently small

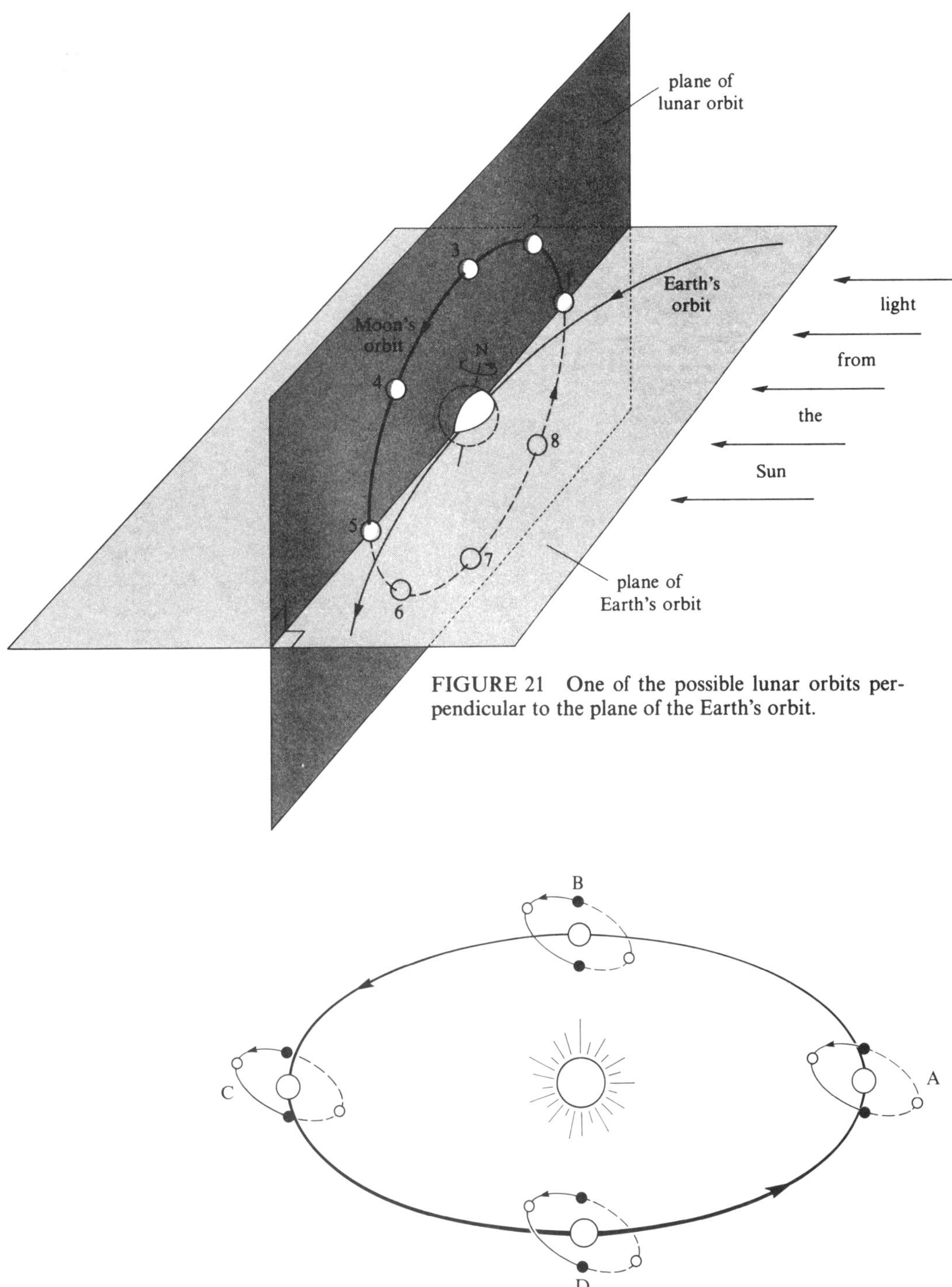

FIGURE 21 One of the possible lunar orbits perpendicular to the plane of the Earth's orbit.

FIGURE 22 What happens if the lunar orbit is slightly inclined to the plane of the Earth's orbit? In each of A–D, the dashed part of the lunar orbit is *below* the plane of the Earth's orbit. Solid circles: Moon in the plane of the Earth's orbit; open circles: Moon above or below the plane of the Earth's orbit (therefore there cannot be an eclipse of the Sun or the Moon).

for the Moon to remain visible from the Northern Hemisphere even at the lowest point of its orbit. This is helped by the fact that the Earth's axis of rotation is also inclined with respect to the plane of the Earth's orbit.

□ Can any eclipses of the Sun or the Moon occur when the Earth is in position A?

■ No. An eclipse can occur only if the Sun, the Moon and the Earth lie in a straight line. This cannot happen in A, because the two points at which the lunar orbit crosses the plane of the Earth's orbit (the two points shown as solid circles at A in the Figure) lie far away from the line that connects the Sun to the Earth. At all other points on the lunar orbit in A, the Moon lies either above or below the plane of the Earth's orbit, and therefore either above or below the straight line joining the Sun and the Earth. Using this type of reasoning, you should be able to appreciate that eclipses can only be seen when the Moon's orbit is in, or around, positions B and D.

So a very slight inclination of the lunar orbit (actually by about 5.2°), just sufficient to bring the Moon above or below the Sun–Earth line in and around positions A and C, brings our model in full agreement with observations. According to our modified model, eclipses cannot happen every month, but they must happen at approximately half-yearly intervals, when the three bodies are in configurations B or D shown in Figure 22 and the Moon is in the plane of the Earth's orbit. The eclipses do not happen on the same dates every year, because the plane of the lunar orbit is not completely fixed. However, this does not alter the argument that they happen twice a year, rather than every month.

Eclipses have considerable emotional attraction for scientists and non-scientists alike, but alas there is not space in this Course to consider them in detail. For the purposes of this Course, you only need to understand why and approximately how often eclipses can happen. Figure 23 should help you to realize why a total eclipse of the Sun is visible only for a short while and from a limited area of the Earth's surface, whereas a total eclipse of the Moon can be seen from any where on the Earth. (Note that in this Figure, for the sake of pictorial clarity, the relative distances between the Sun, the Earth and the Moon are not shown in the correct proportion.)

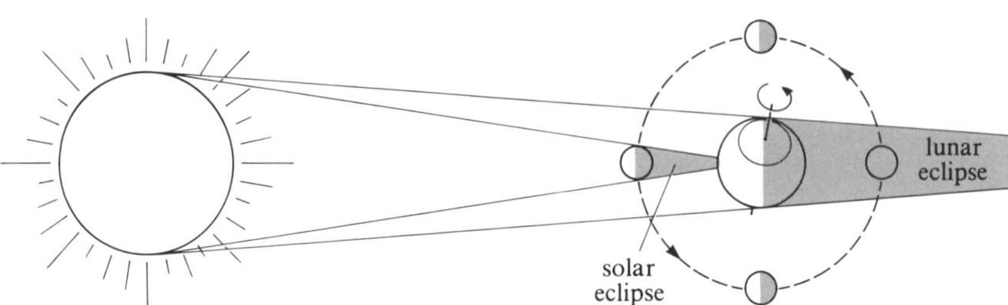

FIGURE 23 Solar and lunar eclipses.

4.2.3 DOES THE MOON SPIN?

There is just one more important observation about the Moon that has not been built into our model so far. How is it that from the Earth we only ever see one and the same side of the Moon? How can this observation be related to the orbital motion of the Moon that had to be introduced into our model in order to account for the cycle of phases? Your first thought might be to say that the Moon does not spin.

□ Why can we rule out straight away the suggestion that the Moon does not spin at all?

■ Because we always see the same face of the Moon. If the Moon did not spin, it would always be oriented the same way in space; for example, a particular feature on the surface would constantly face towards one and the same star. Because the Moon is moving around the Earth, an Earth-bound observer would see the whole surface of the Moon during a lunar month. (Try turning in a circle with an apple held at arm's length, while keeping the stalk of the apple pointing towards, say, the top of a door so that the apple is always oriented the same way in space.)

So, given that the Moon is spinning, what can we deduce about the Moon's axis of rotation and about the period of its spin?

ITQ 13 Select from the key below the *two* statements that, taken together, are capable of explaining why we always see the same face of the Moon from the Earth. Explain why the other statements cannot be correct.

KEY for ITQ 13

A The axis of rotation lies in the plane of the lunar orbit, and always points towards the Earth

B The axis of rotation lies in the plane of the lunar orbit, and its direction is fixed in space

C The axis of rotation is roughly perpendicular to the plane of the lunar orbit

D The period of spin is exactly equal to the orbital period of the Moon

E The period of spin of the Moon is the same as the Earth's period of spin

F The period of spin is the same as the Earth's orbital period

4.2.4 SUMMARIZING THE MOON'S PLACE IN THE MODEL

You are now in a position to review the important features of the Moon's relationship to the Earth and the Sun. If you wish to do so now, try SAQ 5. Alternatively, you might want to return to the question later, as a form of revision.

SAQ 5 There are two sets of statements below. One set lists nine observations relating to the Moon as seen from the Northern Hemisphere of the Earth. The other set lists features (a)–(d) of a model three-body system representing the Sun, the Earth and the Moon. Indicate for each feature (a)–(d) the observations (1–9) that are relevant to each particular feature.

Observations

1 The Moon is visible at different times during the day and/or night.

2 The time at which the Moon is visible is closely related to the shape and size of its bright part (phases).

3 The Moon is much less bright than the Sun and emits a negligible amount of heat.

4 The complete cycle of lunar phases has a period of about 28 solar days.

5 The Moon is visible, albeit at different times, for some part of each solar day (provided that it is not obscured by clouds).

6 The Moon shows the same face to the Earth at all times.

7 The Moon is always of the same apparent size.

8 The apparent size of the Moon is about the same as that of the Sun.

9 Eclipses of the Moon occur relatively rarely (no more than twice a year).

Features

(a) The Moon is a spherical body reflecting the light from the Sun. Choose *two* observations.

(b) The Moon moves around the Earth in a circular orbit. Choose *four* observations.

(c) The plane of the lunar orbit is slightly inclined with respect to the plane of the Earth's orbit. Choose *two* observations.

(d) The Moon spins with a period equal to the period of its orbital motion, the axis of rotation being roughly perpendicular to the plane of the orbit. Choose *two* observations.

4.3 THE EARTH AND THE STARS

During each night the stars appear to move in circles, all in the same direction and always at the same relative distances from each other (the shapes of constellations do not change). In the Northern Hemisphere this rotation takes place about a fixed point, the star Polaris.

Since you already know that the Earth spins about an axis of rotation whose direction is fixed in space (Sections 4.1.2 and 4.1.3) and that this spin explains the apparent circular path of the Sun, it is only natural to assume that the apparent rotation of the stars is also explained by the spin of the Earth.

□ Can you think of a simple experiment by which this assumption can be tested?

■ You could, for example, set a photographic camera on a firm support and point it so that Polaris was at the centre of the field. Then if you were to leave the shutter open for a sufficiently long time — at least four hours — you would obtain a photograph looking something like Figure 24. The lengths of the curved tracks depend on the time of exposure.

ITQ 14 Select one of the more distinct tracks in Figure 24 and estimate approximately what fraction of a full circle it represents. From this, decide what exposure would, in your view, produce such a picture: 2 hours, 4 hours, 8 hours, 12 hours, or 24 hours.

FIGURE 24 A photograph of the night sky, with Polaris at the centre, taken with a long time-exposure to show the motion of the stars. (Note that in order to increase the clarity of this copy, some bright tracks were emphasized and many weaker tracks were suppressed.)

The observations that *all* star tracks on the photograph are concentric circular segments and that each segment is *the same fraction* of a full circle, support the assumption that the apparent motion of stars is due to the spinning of the Earth. If that is so, then the only possible explanation for the fact that Polaris never moves is that *Polaris lies exactly in the direction of the Earth's axis of rotation.*

We can now combine this conclusion with the previous result derived in Section 4.1.3. In order to explain the seasonal cycle within one year and its local variations at different parts of the Earth's surface, it was necessary to

FIGURE 25 The Earth's axis of rotation points permanently to Polaris.

accept that the axis of rotation is tilted with respect to the plane of the orbit. In fact, for the best agreement with all observations the axis must be at an angle of about 67° to the plane of the orbit (Figure 25). Now we know that this direction is also the direction to Polaris.

SUMMARY OF SECTION 4

In this Section, we have developed a model of the relationship between the Earth, the Sun and the Moon, in order to account for the experimental observations discussed in Section 3. Here is a summary of the model.

1 The Sun, the Earth and the Moon are spherical bodies.

2 The Sun radiates light and heat evenly from every part of its surface (hence spin would not be noticeable, even if it exists).

3 The Earth spins about its axis of rotation and this accounts for the day–night cycle.

4 The Earth orbits around the Sun in a circle and one complete circle corresponds to one complete seasonal cycle (one year).

5 The Earth's axis of rotation is inclined at about 67° to the plane of its orbit, remains fixed in space and points towards the star Polaris. The tilt of the axis of rotation accounts for seasonal variations.

6 The Moon reflects sunshine and it moves around the Earth in a circular orbit. During one orbital period (one month), the Moon goes through a cycle of phases; these can be explained by its relative positions with respect to the Sun and the Earth.

7 The plane of the lunar orbit is slightly inclined with respect to the plane of the Earth's orbit. This explains the times at which the Moon is observed above the horizon as well as the frequency of eclipses.

8 The Moon spins about its axis of rotation, which is perpendicular to the plane of its orbit, and the period of its spin is exactly equal to the period of its orbital motion around the Earth. This explains why the same face of the Moon is seen at all times on Earth.

5 TV NOTES: THE PLANET EARTH — A SCIENTIFIC MODEL

It is not very easy to visualize the three-dimensional relationship between the various bodies that orbit the Sun. In such a situation, it is helpful to construct a three-dimensional model of the system; we can then 'look down' on the model and explore its behaviour from an external viewpoint. That is what is done in this TV programme. As you will see, this is a particularly useful approach to adopt in trying to understand the Solar System, as it helps us to escape from our Earth-bound viewpoint (or 'frame of reference') and see, in our mind's eye, the workings of the Solar System from outer space. And this, in turn, helps us to understand why things appear the way they do when observed from our spinning, orbiting Earth. In particular, this 'double viewpoint' approach is used in the programme to explain:

(a) the (at first sight) peculiar behaviour of a Foucault pendulum;

(b) the phases of the Moon;

(c) the infrequency of lunar eclipses;

(d) the 'retrograde' motions of planets.

When viewed from the 'frame of reference' of a rotating roundabout, the motion of a swinging pendulum (attached to the roundabout) appears very strange; the plane-of-swing of the pendulum appears to be continuously changing. This point is demonstrated in the opening sequence of the programme. However, when the pendulum is viewed from a *stationary* frame of reference relative to the roundabout, it can be seen that the plane-of-swing of the pendulum is fixed. The apparent change of the plane-of-swing (as seen from the roundabout) is a consequence of the roundabout's rotation. This is the basic principle underlying the operation of a Foucault pendulum. The programme demonstrates that the plane-of-swing of a pendulum suspended directly over the Earth's North Pole remains fixed *with respect to the stars*. If the Earth is spinning, an observer attached to the Earth will see an apparent rotation of the plane-of-swing. Conversely, if the plane-of-swing of a suspended pendulum appears to rotate, then we can deduce that the Earth is spinning. In the programme, we show you a modern reconstruction of this classic experiment (first performed in 1851 by the French physicist Leon Foucault). Our pendulum was suspended from the dome of St Paul's Cathedral in London (Figure 26). The experiment is not as simple to set up as you might think. We had to take all kinds of precautions to ensure that the rotation of the plane-of-swing was genuinely caused by the Earth's spinning motion, and not by bias in the suspension, by 'twist' in the suspension cable, or by a small sideways motion imparted to the pendulum bob at the time of release.

We show that, in London, the plane-of-swing rotates at the rate of about 12° per hour. This means that it takes (360°/12°) hours = 30 hours for a rotation through a complete turn of 360°. (The rate is exactly 15° per hour — 24 hours for 360° — at either the North or South Pole and decreases to zero at the Equator.)

The programme then goes on to examine the phases of the Moon. By building a model of the Earth–Moon system, and by simulating the sunlight with a distant arc lamp in the studio, we are able to compare the view of the

FIGURE 26 Performing the Foucault experiment in St Paul's Cathedral; Mike Pentz, former Dean of Science at the Open University, is on the left, and the Dean of St Paul's on the right.

'Moon' (as seen from a camera positioned on the 'Earth'), with the view of the whole system as seen from a camera positioned in 'outer space'. It is then possible to see that the full-Moon and new-Moon phases occur when the Sun, Earth and Moon are all aligned in the same plane, whereas half-Moon (that is, the first or last quarter) occurs when the angle between the Sun and the Moon (as measured from the Earth) is a right angle (90°).

However, this simple method leads to the prediction that a lunar eclipse should occur every month. Lunar eclipses are not nearly as frequent as this, so the model needs to be modified. We show that the inconsistency can be removed by *inclining* the plane of the Moon's orbit about the Earth. Lunar eclipse positions are much less frequent according to such a model.

Finally we move outwards from the Moon, to the planets of the Solar System. And again we use the 'double viewpoint' approach to help explain the phenomenon of retrograde motion.

6 CONCLUDING REMARKS

If you were now to ask yourself what you have learnt from this Unit, the answer would depend far more on your previous knowledge than on what you have read here. Perhaps you found no new information at all. If so, remember that the communication of information was not the main purpose of the exercise. We wanted, above all, to convey to you something of the way in which science develops. In this Unit, you have repeatedly seen examples of the process first described in Section 1.2. By working on observations or experience, the human imagination creates a trial explanation or theory. Sometimes, it even provides more than one theory, and the different possibilities then compete for our allegiance. Let us recall some examples.

You saw the process in action in Section 2 with the pancake and spherical models of the Earth, and in Section 4.1 with the four options for the relative motion of the Earth and Sun. It recurred again in Section 4.1.3 with attempts to explain the seasonal cycle, either by modification of the Earth's orbit in our model or by tilting the Earth's axis of rotation. In Section 4.2, imagination first offered a choice between a Moon moving alongside or in orbit around the Earth, and then between various orientations of the Moon's orbit with respect to the Earth's orbital plane. All these different theories were essentially descriptions of models of the type first mentioned in Section 1.2.1.

What happened after these different theories were introduced? In all cases, further observations or experiences were drawn into the discussion. Sometimes these flatly contradicted deductions that could be made from some of the theories, but were consistent with conclusions that could be drawn from others. Thus some of the theories could be falsified and others corroborated, as described in Section 1.2. Progressive falsification of the competing models continued through the Unit until, at this stage, just one has survived.

It is important to note that the simple model described in the Summary of Section 4 is not consistent with *all* the relevant observations that have ever been made, only with those we have introduced so far. As soon as one takes a more careful look, using accurate measuring instruments, the model has to be refined. Thus the Earth is not perfectly spherical, although this is not apparent from mere inspection of the photograph at the beginning of the Unit. Likewise the Earth's orbit is not exactly circular, although the deviation from a circle would not be noticeable if the orbit were reduced to the size of this page.

By now, however, none of this should surprise you. In Section 1.2.1, you were told that models are often constructed for narrow purposes, and this model has coped adequately with all the observations that have concerned us until now. And as was stressed in Section 1.2.1, science does not claim that its theories are true; they are retained provisionally until new evidence turns up to prove them inadequate.

SOLAR SYSTEM

Finally, a few words on Section 4.1.4, where a choice was made between two competing models, even though neither of them had been falsified by experience: the heliocentric model was preferred to the geocentric one. This choice was made because in the geocentric model the observed retrograde loops in the planetary paths would mean real reversals of the direction of motion, for no apparent reason. The heliocentric model, by contrast, explains this observation by a very simple combination of two relative circular motions. Now in science, *one does not choose a more complicated explanation when a simpler one does the job equally well.* So the heliocentric model was preferred for its greater simplicity and elegance, or, dare we say it, for its beauty. It is illustrated in Figure 27, which shows schematically the bodies orbiting the Sun — the **Solar System**.

In a sense, therefore, the choice was an act of faith. And subsequently, that faith was fully justified. By putting the Sun at its centre, the model facilitated the huge imaginative leaps to the laws of planetary motion and gravitation. Those leaps are associated particularly with the names of Johann Kepler and Isaac Newton.

To understand their achievements, you need to delve further into the nature of science. Throughout this Unit, you have not used a single formula or equation. And yet, in a very broad sense, the derivation of the heliocentric model was mathematical, in that you were making choices between different options, and eliminating wrong ones by combining reasoning with new observations. This approach is fine as far as it goes, but it does not go far enough.

Our curiosity is not satisfied by knowing that the Earth goes round the Sun, and the Moon around the Earth. What about the sizes of the three bodies? How far apart are they? How fast do they move with respect to one another? These are *quantitative* questions. To answer them, one must make *measurements*, and begin to manipulate mathematical formulae. You will take this step in Unit 2, and it will give you the necessary background for you to understand the principles that explain the motions of *all* celestial bodies.

OBJECTIVES FOR UNIT 1

When you have completed this Unit, you should be able to:

1 Explain the meaning of, and use correctly, all the terms identified by bold type in the text. (These terms are also 'flagged' in the margins at the top of appropriate left-hand pages.)

2 Distinguish statements that are scientific from those that are not, according to the falsifiability criterion. (*SAQ 1*)

3 Describe what is meant by a periodic process, and be able to work out the period, given the appropriate data. (*ITQs 2–5*)

4 Describe the motion of the Earth with respect to the Sun. (*ITQs 7–9; SAQs 2–4*)

5 Describe the motion of the Moon with respect to the Earth and the Sun. (*ITQs 10–13*)

6 Explain local variations in the seasonal cycle. (*ITQs 7–9; SAQ 2*)

7 Explain the phases of the Moon. (*ITQs 10–12*)

8 Explain the apparent motion of the stars. (*ITQ 14*)

9 Correlate systematic observations (such as the sequence of seasons, lunar phases, etc.) with the features of a model. (*ITQs 1 and 7–13; SAQs 2–5*)

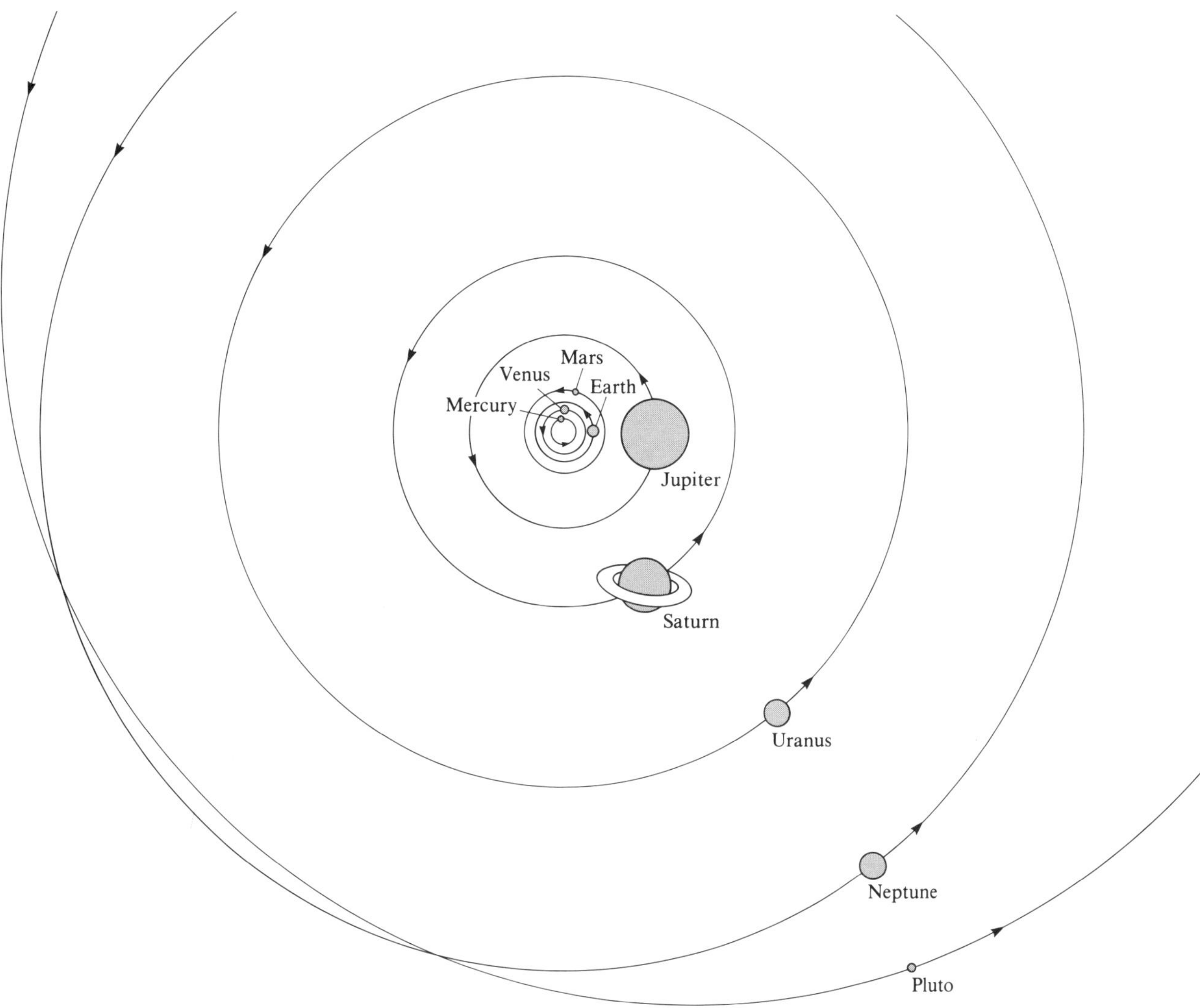

FIGURE 27 Schematic diagram of the Solar System. The diameters of the planets are shown in approximately the correct proportion to each other, as are the sizes of the orbits. However, the sizes of the orbits are *not* in correct proportion to the diameters of the planets. (If the orbits had been drawn on the same scale as the sizes of the planets, this page would have had to be more than 1 kilometre wide!) The diameter of the Sun is about ten times that of Jupiter, so the Sun could not be shown at the centre of this Figure; on the scale we are using for the planets, it would have covered up most of the Figure!

ITQ ANSWERS AND COMMENTS

ITQ 1 (a) It is impossible to tell which sort of curvature the Earth has.

The shape of the surface cannot be determined from the evidence given in the question. *One* observation cannot distinguish between the various possible shapes of a curved surface.

(b) That the distance at which the ship disappears is the same, wherever it starts from and whatever the direction in which it sails.

For most of the surfaces shown in Figure 6, the rate of curvature is different at different points of observation and in different directions. On the surface of a cylinder there is even one direction along which there is no curvature at all, and an infinite number of directions of no curvature can be found on a cone (e.g. looking from the tip). A sphere, on the other hand, can be described as having the same curvature in all directions at any point on its surface. Thus, to show that the surface of the ocean is spherical, one would have to show that a ship departing from one point in *different* directions always disappears at the *same* distance. Moreover, this distance of disappearance would have to be the same in all directions from all points of departure.

ITQ 2 (a) 2 seconds; (b) 0.2 second; (c) 0.2 second.

(a) Because the disc completes 5 revolutions in 1 second, to complete 10 revolutions it must take

$2 \times 1 \text{ second} = 2 \text{ seconds}$

(b) Because the disc completes 5 revolutions in 1 second, it must complete 1 revolution in

$1 \text{ second}/5 = 0.2 \text{ second}$

(c) The period of rotation of the disc is defined to be the time it takes to complete one revolution — 0.2 second, as you saw in part (b)!

ITQ 3 Shortening the pendulum reduces the time.

You should have found that the shorter the pendulum, the faster it swings (i.e. the more complete swings it executes in a given time). The time taken to execute 10 swings for a 1 metre pendulum is about 20 seconds. We shall not give you the times for the other two pendulums — we may ask you about them in one of the Unit's CMA questions!

ITQ 4 The shorter the pendulum, the shorter is its period.

Remember the period is defined as the time taken to execute one complete sequence of the periodic motion, in this case one complete swing. Hence, from the answer to ITQ 3, the period of the 1 metre pendulum is about 20 seconds/10 = 2 seconds.

By the way, you may have noticed that, for each pendulum, each successive swing has a slightly smaller amplitude (the amplitude is the distance between one of the extreme positions and the vertical). Yet, for each pendulum, there should have been no noticeable difference in the period of the successive swings.

ITQ 5 Before you performed the experiment, you may well have guessed that a pendulum with a heavier bob would swing more slowly (i.e. have a longer period) than a pendulum of the same length with a lighter bob. Having performed the experiment, you should have come to the conclusion that the period does not depend on the mass. It should have taken your heavy-bob pendulum about the same time to execute 10 swings as it took the light-bob pendulum of the same length. This may well be counter-intuitive, yet it has very important consequences. As you will see in the TV programme associated with this Unit, it enables us to make the bob of a Foucault pendulum (designed to measure the spin of the Earth about its axis of rotation) very heavy, without affecting the pendulum's period. The advantage of this is that it takes a very much longer time for this massive swinging pendulum to come to rest than would be the case with a lighter pendulum.

ITQ 6 Provided that you did not, either deliberately or accidentally, interfere with the swinging of the pendulum, the plane of swing should *not* have changed. Thus you can conclude that the plane of the pendulum's swing is constant, regardless of the rotation of the point from which it is suspended.

ITQ 7 Going through the elliptical orbit in Figure 13a, you should expect the Earth to experience two complete seasonal cycles within one orbital period. There are two positions, 2 and 4, where the separation of the Sun and Earth is least (these would presumably correspond to summer) and two positions, 1 and 3, with maximum separation (winter). Spring would occur between positions 1 and 2, and also between positions 3 and 4, while autumn would occur between 2 and 3, and 4 and 1. The orbit of Figure 13b, which is circular but displaced so that the Sun is not at the centre of the circle, would lead to one cycle of seasons per period. Presumably, position 1 would be winter and position 3 summer, with a gradual change to spring at position 2 and autumn at 4. Either model *could* be correct.

ITQ 8 In and around position 4 in Figure 14a, the observer in Britain would see the Sun all the time. There would be no day–night cycle but continuous daylight. On the other hand, around position 2, Britain would have a long period of continuous darkness (facing away from the Sun). In positions 1 and 3, Britain would experience a regular day–night cycle, but so would all other places all over the Earth. The pattern would be identical everywhere; there would be *no seasonal difference between Britain and Australia.*

Note, incidentally, that the angle at which we draw the Earth's axis of rotation *within the plane of the paper* (i.e. within the plane of the orbit) is irrelevant. For instance,

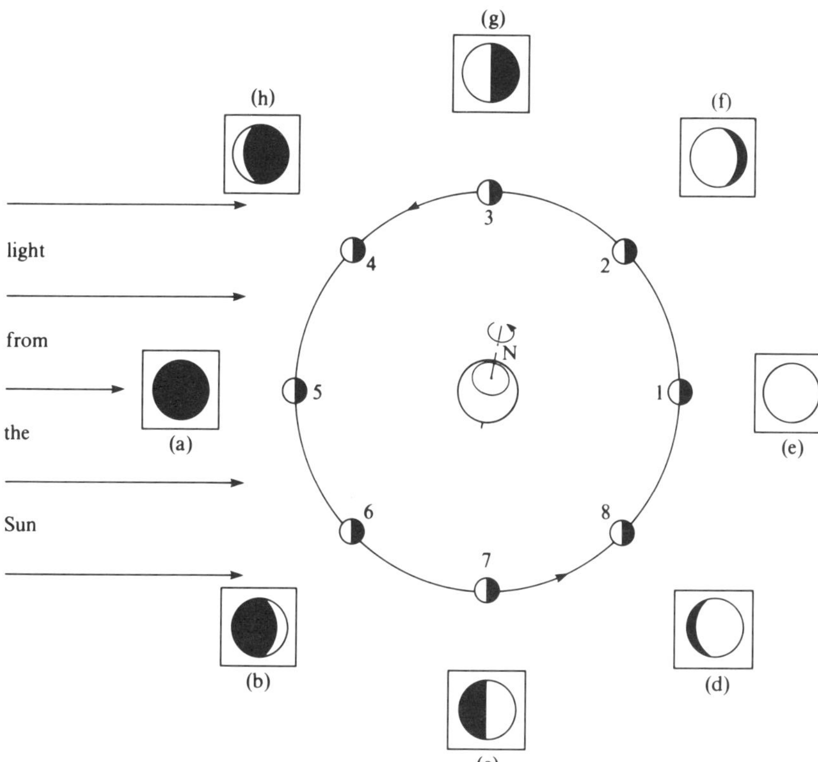

FIGURE 28 Phases of the Moon (for ITQs 10 and 11). Refer to Figure 20 for (a)–(h). *Note*: If the plane of the lunar orbit were *identical* with the Earth's orbital plane (as assumed in Figure 19), the Moon would be eclipsed *every time* it reached position 1. This does not happen; most of the time we see a fully illuminated Moon (e) at position 1.

had we drawn the axis across the page in all the positions 1 to 4 (rather than up and down the page), this would simply have been equivalent to turning the whole Figure through 90°. Try it!

ITQ 9 Situation 4 in Figure 14a is characterized by the Sun facing the Northern Hemisphere of the Earth. There is an identical relative configuration of the Earth and the Sun in Figure 14b, where the Sun is at the top of the Figure. Similarly for each numbered position in Figure 14a you can find an identical configuration in Figure 14b. Thus, an observer on the Earth could not tell which of the two bodies is orbiting — all observations would be the same in both cases.

ITQ 10 (i) 6; (ii) 1; (iii) 3; (iv) 2.

Perhaps the easiest way to approach this question is as follows. First, you should realize that the lunar hemisphere facing *away from* the Sun (i.e. the right-hand hemisphere in all positions in Figure 19) will always be dark. Try shading this in on Figure 19 before proceeding.

Now you have to imagine what an observer in the Earth's Northern Hemisphere would see when looking directly towards one of the numbered positions of the Moon. You can perhaps help your visualization a bit here by drawing two diverging lines away from the point on the Earth that represents the observer's position, to the extremities of the lunar disc at the orbital position under consideration. If you do this for position 3, for instance, you should be able to convince yourself that the observer sees the right-hand side as dark, i.e. the Moon is in the last quarter. Similarly, for position 6, the observer sees darkness covering almost all of the circle, with only a small crescent of light on the right-hand side *as viewed from the Earth*, i.e. position 6 must be the waxing crescent. If you had trouble with answering ITQ 10, have another try now at answering (ii) and (iv).

ITQ 11 Using exactly the same sort of 'visualization aids' as in ITQ 10, you should be able to work out that:

position 4 is a waning crescent (h);

position 5 is the new Moon (a) (eclipsing the Sun);

position 7 is the first quarter (c);

position 8 is the waxing gibbous phase (d).

All the Moon's phases discussed in this ITQ and in ITQ 10 are shown in Figure 28.

ITQ 12 (i) 7; (ii) 2; (iii) 5; and (iv) 8.

The way to tackle this question is to identify first the position of an observer on Earth experiencing dawn, noon, dusk and midnight. Convince yourself that these experiences correspond, respectively, to an observer looking towards position 3, position 5, position 7, and position 1. Can you now see that if the Moon is in position 3 it will be visible (i.e. above the horizon) from roughly midnight through until noon? Thus, to be visible from noon until midnight (condition (a) in the question), the Moon must be in position 7. Try reasoning out the rest for yourself.

ITQ 13 C and D.

The Moon spins about an axis of rotation which is roughly perpendicular to the plane of the lunar orbit, and with a period of spin exactly equal to the Moon's orbital period. In this way, the changing view of the Moon that we would expect as it circles the Earth would be *exactly* counteracted by the spinning of the Moon around its own axis of rotation.

If statement A were true, the Earth-bound observer *would* always see the same half of the lunar sphere, but all the features away from the lunar pole would rotate in circles. Anyway, as we pointed out when discussing the Foucault pendulum, we would expect the axis of rotation of a freely spinning body to be fixed in space. Here it turns so that it always points towards the Earth — very odd behaviour!

Statement B overcomes the 'Foucault pendulum' objection cited above, but now different faces of the Moon will point towards the Earth as the Moon executes one orbit. That is, the axis of rotation will point towards the Earth only twice in one orbit (once for the 'north' lunar pole, and once for the 'south' one).

If statement E were true, then during the 12 hours or so when the full Moon is continuously visible, its surface features could not remain static. The Moon would complete about half a turn during this time.

Finally, if the Moon's period of spin were about one year (statement F), then during one lunar month (the time the Moon takes to complete one orbit of the Earth) a fixed surface feature would have turned by about 1/12 of a full circle. Thus the parts of the Moon visible at two consecutive full-Moon phases would be different.

ITQ 14 8 hours.

Because the tracks are each about one-third of a full circle and a full circle would be completed in 24 hours, the exposure time must have been (24 hours)/3 = 8 hours. Exposures as long as 12 hours would be possible only in winter and would produce half circles. An exposure of 24 hours would ruin your picture because of the daylight (unless taken during a polar night).

SAQ ANSWERS AND COMMENTS

SAQ 1 Although there is always room for argument, we suggest that, according to the falsifiability criterion, statements (a), (b) and (d) are scientific, and statements (c), (e) and (f) are not.

Statement (a) could conceivably be falsified by production of a white raven, and statement (b) has probably already been falsified by your own experience. This latter example shows that, if the argument of Section 1.2 is accepted, it is quite unnecessary for a scientific statement to be true. Statement (d) could be falsified by interplanetary reconnaissance.

The other statements are not scientific because they contain let-outs that can be appealed to when the statements are not corroborated. With statement (c), a committed believer can always claim that life exists where we have not yet looked; with statement (e), that capitalism's doom, though certain, is not yet come; and with statement (f), that the ceremony cannot have been correctly performed. Notice, however, that all of the three statements that we have classified as not scientific, *could* be true. Indeed, one day, extra-terrestrial life forms may be found, capitalism may be overthrown and the Devil may be conjured up, corroborating the statements; but there is no way they could be *proved false.*

SAQ 2 Your completed Table 3 should be as shown in Table 4.

TABLE 4 For the answer to SAQ 2

Location of observer	Configuration 1	2	3	4
Northern Hemisphere	winter	spring	summer	autumn
Southern Hemisphere	summer	autumn	winter	spring

In position 1, the Northern Hemisphere is tilted away from the Sun, so it will receive less light and heat and the Sun will appear to culminate low above the horizon (winter). By contrast, the Southern Hemisphere will be inclined towards the Sun — longer days, the Sun culminating high above the horizon, more light and heat (summer). In position 3, the relative inclination of the two hemispheres with respect to the Sun is reversed. In positions 2 and 4, both hemispheres go through the transitory mild periods.

SAQ 3 The only thing that matters in the determination of the season is the relative inclination of the hemisphere towards or away from the Sun. Thus position 1 in Figure 15a is the same as position 3 in Figure 15b, where the Sun is to the left of the Earth (Northern Hemisphere tilted away from the Sun — winter). Similarly, a(3) is equivalent to b(1) (summer), a(2) to b(4) (spring), and a(4) to b(2) (autumn).

SAQ 4 The final two items in the Table should be as shown below.

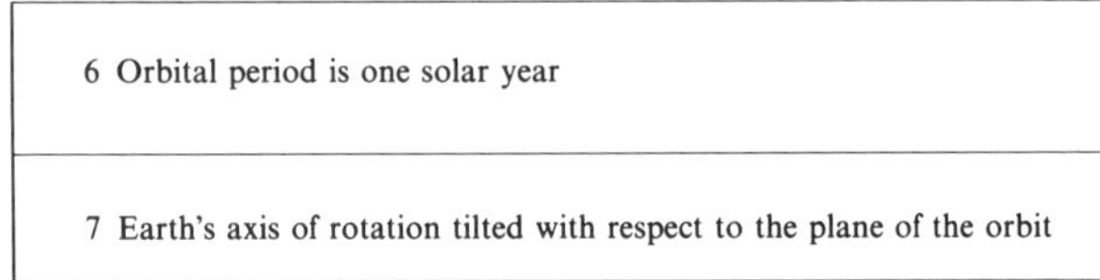

6 Orbital period is one solar year
7 Earth's axis of rotation tilted with respect to the plane of the orbit

SAQ 5 (a) Observations 2 and 3; (b) observations 1, 2, 4 and 7; (c) observations 5 and 9; (d) observations 4 and 6.

All observations 1–9 are consistent with a model that has features (a)–(d). However, some of the observations are particularly relevant to some features. Note that observation 8, although interesting, does not relate to any of the features (a)–(d). It will, however, assume a great significance in Unit 2, when you will be considering the sizes of celestial bodies and their distances from the Earth.

INDEX FOR UNIT 1

Entries and page numbers in bold type refer to key words.

ACKNOWLEDGEMENTS

Grateful acknowledgement is made to the following for permission to reproduce Figures in this Unit.

Figure 1 NASA; *Figure 2* Syndication International; *Figure 3* Tate Gallery, London.

THE OPEN UNIVERSITY
A SCIENCE FOUNDATION COURSE

UNIT 2 MEASURING THE SOLAR SYSTEM

STUDY GUIDE

This Unit has three major components: the text, an experiment, and the TV programme 'Measuring—the Earth and the Moon'.

Although this Unit is ostensibly about the Earth and its position in the Solar System, it is also designed to introduce you to a number of mathematical skills that you'll need to use many times during the Course. Consequently, the text of the Unit is laid out very much in the manner of a workbook: you are asked to fill in blanks, complete Tables, draw graphs and, of course, answer lots of questions. In fact, the motto of this Unit might well be 'learning by doing'.

In order to give you plenty of practice in mastering a new skill, its first introduction in the text is usually followed immediately by one or more questions. All these questions are labelled as ITQs, although, since they also enable you to test yourself as you go along, they actually serve as ITQs and SAQs rolled into one. For this reason, there are no separate SAQs in this Unit.

The main thrust of the story also includes an important experiment, in which you will measure the distance to the Moon. This experiment, and the activities leading up to it, are to be found in Section 3.4; if you want a quick preview of the apparatus you will need to assemble (some of which you must provide yourself), look at Section 3.4.1. Although you won't be able to *analyse* the results of your experiment until you have read to the end of Section 3.4, you can do the *practical* part of the work whenever this is convenient. You should therefore **seize the opportunity of the first cloud-free night to make the measurements**; it isn't essential that the Moon be full, but the experiment is easiest to perform within a day or two either side of a full Moon.

The TV programme, 'Measuring—the Earth and the Moon' is associated mainly with Section 3. Notes on the TV programme are given at the end of the text. In order to derive most benefit from this programme, you must have read to at least the end of Section 3.4 before watching it. Part of the programme is used to explain how to make an important correction to one of the measurements you take in the experiment. However, you should not postpone doing the experiment until after you have seen the programme—do the experiment at the first available opportunity. The TV programme simply describes how to apply a correction to your data to make the result more accurate—it neither invalidates your measurements nor provides any extra information on the experimental procedure.

1 INTRODUCTION

This Unit is about measurement—about the assumptions made, the reasoning used and the limitations involved, whenever you make a measurement. In order to understand this topic, you must do more than simply listen to what other people have to tell you. You must learn to make measurements for yourself, and you must work to understand why other scientists have chosen to make the measurements they have made. That is why, during the course of this Unit, you will be asked to carry out your own experiment, compile Tables of results, and criticize constructively other people's measurements. Rather than ask you to learn these techniques and skills as an end in themselves, the Course Team has tried to weave the techniques into the story of the early discovery of the 'rules' governing the Solar System. Because you will be *using* scientific skills in the context of this fascinating story, and have the opportunity to practise them as you go along, you should find yourself encouraged and motivated to acquire the necessary expertise.

Many of these skills require mathematics. Making and interpreting measurements often involves using mathematical reasoning. For instance, a *direct* measurement of the distance between the Earth and the Sun is not possible. So an *indirect* approach must be adopted. This may well mean measuring angles (so you must know about angular measure) and then deducing the distance using a 'triangulation' method (so you must know something about the properties of triangles). These mathematical tools are introduced as and when they are needed. You may find that the explanation of them in this Unit is quite adequate for you. If so, you probably won't find the Unit too difficult. But if you're not very familiar with the mathematics, please make sure that you take the advice, given at various places throughout the text, to refer to the explanations and examples in the relevant Sections of *Into Science*.*

You may be wondering why it is necessary to bother with this *quantitative* aspect of science at all. Why not just concentrate on the ideas and concepts of science? Well, in some areas we do just that. Most of Unit 1, for example, was *qualitative* in approach, and there was plenty of 'good science' in that. But most science can't stop at the qualitative level. Sooner or later, two or three qualitative explanations of a phenomenon will come into conflict. And if science is to be a little more objective than the 'my-explanation-is-as-good-as-yours' argument will allow, there has to be some way of choosing between these rival explanations. That is where measurement becomes important.

Of course, the desire to choose between one scientific model and another does not have to be the sole motivation for making measurements! Indeed, people have long realized that there are more down-to-earth reasons why they should be able to measure things. How else, for instance, could they know the quantity of grain they had for sale, or the distance they would have to carry it to market, or the time it would take to walk there? No, the need to measure things existed long before the need to differentiate between contending theories about the Solar System.

One consequence is that we have inherited a variety of measurement standards. Naturally, these standards have evolved over the centuries, from the rather crude and variable measures of primitive people, to the highly precise measurement standards used in modern science. We no longer measure distances in terms of a man's stride: we now have length standards that are accurate to 1 inch in 10 000 miles. And the immensely impressive 'stone calendar' at Stonehenge (Figure 1) has been superseded by an atomic clock which operates with an accuracy of 1 second in 30 000 years. Indeed, this evolution of measurement standards has been absolutely essential for the development of science. For without a system of well-defined units and precise, reproducible standards, scientists would find it impossible to com-

* The Open University (1993) *Into Science*, The Open University.

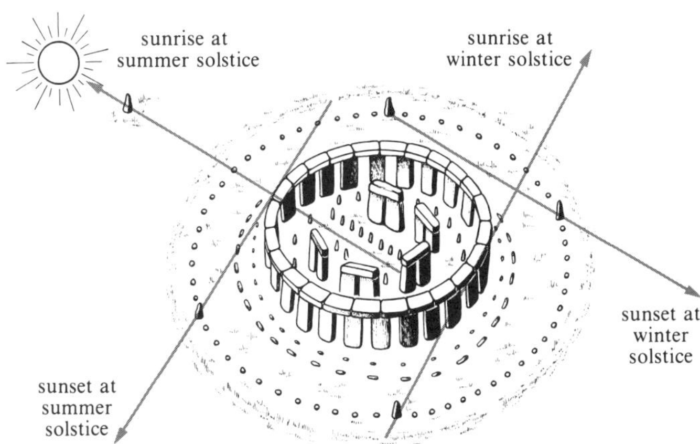

FIGURE 1 Stonehenge. The alignment of various stones is related to the position of the rising and setting Sun on the days of the summer and winter solstice. Sunrise on the summer solstice appears to have been of particular importance; the alignment is such that this sunrise has to be viewed from the altar stone, inside the monument.

municate their findings to one another unambiguously. It is therefore essential that you become acquainted with the international system of units and standards in present-day use. That is why the whole of Section 2 is devoted to a discussion of these standards. However, you should not try to memorize all the details given in this Section. Instead, you should view it as providing background information to help you put your understanding of the measurement techniques described in the rest of the Course into the context of modern science and technology.

The story of the measurement of the Solar System (which takes up most of the remainder of the Unit) is divided into three parts. The first part (Section 3) describes how the early Greek astronomers estimated the sizes and distances involved in our Sun–Earth–Moon system. Section 4 is basically concerned with measurements involving the other planets of the Solar System, and in particular with the pioneering work of Copernicus, Tycho Brahe, and Kepler. The culmination of this work was the formulation, by Kepler, of his 'laws of planetary motion'—three laws that succinctly summarize the *regularity* of the motion of the planets in the Solar System. An epilogue to this story of Kepler's discovery of his three laws is provided in Section 5. Galileo, using the newly invented telescope, discovered four moons circulating around Jupiter. Kepler's laws were applicable to the planets orbiting the Sun. Could they also be applied to the moons orbiting Jupiter? If so, what was the 'mechanism' behind these laws? You will not find the answer to this last question in Unit 2. The 'explanation' for Kepler's laws was provided by Newton, and Newton's contribution to scientific thought is the main focus of Unit 3.

2 MEASUREMENT: UNITS AND STANDARDS

2.1 SETTING THE STANDARD

Measurement can usually be reduced to a comparison of something of interest with some agreed standard. For instance, if you were to take the width of this book as a unit of length, the lengths of all other objects could be related to this width. One object may be twice as long as the width of the book (i.e. two 'book-widths'); a second may be three-and-a-half times as long (3.5 book-widths); another may be a quarter as long (0.25 book-widths). Once the unit is established, measurement becomes simply a matter of counting. So, if you were told that the Walton Hall boiler-house is 60 book-widths high, you would have some idea of how tall it is, even though you may never have seen it! But note that it was important for you to know what the unit of measurement was; furthermore, you even had access to a 'copy' of it. Someone who did not know what the unit of measurement was, or who did not have access to the 'standard' book-width, would have no idea what was meant by a height of 60 units. We need standards, and these standards must be known by everyone with whom we wish to communicate. But what standards should we choose?

SI UNITS

METRE

2.2 STANDARDS OF LENGTH

Early standards of length tended to be related to the size of the human body. A 'cubit' was the distance between the elbow and the tip of the middle finger. The width of a man's hand was used to express the height of a horse in 'hands'. The width of the thumb became an 'inch', the length of the foot became a 'foot'. When measuring cloth, it was convenient to define the distance from the nose to the end of the middle finger when the arm was outstretched as a 'yard'.

These 'personal' standards, though conveniently portable, can, however, vary considerably from person to person. Get two drapers together, for instance—a tall one and a short one—and, with this definition of the yard, one of them sells you 'cheaper' cloth! Clearly, we need a greater degree of standardization than this. So what do we do? We invent the *yardstick*. Copies of this stick can then be sent all over the country, with the result that a yard of cloth in Edinburgh is more or less the same length as a yard of cloth in London. But what do we mean by 'more or less?' To within $\frac{1}{10}$ inch? Nobody is likely to grumble about that sort of uncertainty if they are buying cloth; but in science we may well want to measure to smaller subdivisions than that. So the 'precision' of our standard must improve. The ultimate limit to the accuracy with which we can make a measurement (no matter how good the equipment) will always be determined by the uncertainty in the standard. It is no good giving a measurement to an accuracy of $\frac{1}{100}$ inch, if the agreement as to what constitutes a standard yard is only good to $\frac{1}{10}$ inch. (Mind you, once you have found a way of measuring to an accuracy greater than that to which the standard is defined, the onus is on you to petition for a new standard based on your improved technique!) Hence, as scientists have sought to make more and more accurate measurements, so they have also had to devise more and more precise measurement standards.

Nowadays, there is also a further requirement: our measurement standards should be international. Clearly, this was unimportant when different communities did not interact with one another. Today, however, if scientists in Europe are to understand the measurements of scientists in the USA, the USSR, or Japan, they must be acquainted with one another's standards of measurement. Better still, they should all use the same standards. In 1960 it was formally agreed to standardize according to the 'Système International d'Unités'. This system, usually abbreviated as **SI units**, is now in universal use in the scientific community and is the one with which you will be primarily concerned in this Course. The SI system is *metric*, i.e. based upon the **metre** as the standard unit of length.

The history of the metric system dates back to Napoleonic France, where the metre was originally defined as one ten-millionth of the distance from the Equator to the North Pole along a meridian passing through Dunkirk and Barcelona, and hence also passing very close to Paris (Figure 2). This choice of standard was a very 'safe' one—the standard could not be lost! But it could hardly be called practical. So in 1889 it was officially decided to define the metre as the distance between two parallel marks inscribed on a particular bar made of the metal alloy platinum–iridium. (This alloy was chosen because of its exceptional hardness and its resistance to corrosion.) Furthermore, to ensure the reproducibility of measurements of this length, the bar had to be kept under specific conditions. For instance, it had to be supported in a particular way, so as to minimize deformations, and it had to be kept at the temperature of melting ice so as to prevent the expansion or contraction that would occur if the temperature were allowed to vary. This *standard metre* still exists. It is housed in the International Bureau of Weights and Measures in Sèvres, near Paris. Copies of this bar, i.e. *secondary standards*, have been made and distributed to national standards offices throughout the world.

Even this, however, was not completely satisfactory. For although the length of an object could be compared with the standard metre to a precision of about two parts in ten million (by using a high-powered microscope

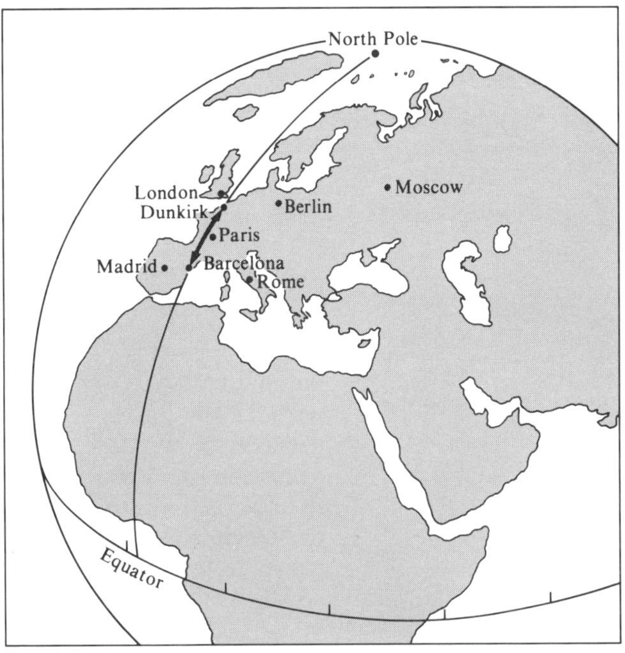

FIGURE 2 The metre was originally defined as one ten-millionth of the distance from the North Pole to the Equator. A team of French surveyors was engaged to measure the distance between Dunkirk and Barcelona. The length of the full quadrant was then determined from astronomical measurements of latitude.

to view the finely inscribed marks on the metre bar), this precision was still inadequate for some scientific purposes. (Don't forget, the precision to which we know the standard must be better than the *most* accurate measurement we wish to make.) In addition, making comparisons with a bar that had to be kept under specific conditions in a standards laboratory was still inconvenient. What was required was a standard that anyone (or at least, any scientist) could have access to in any laboratory, a standard that did not require copies to be made (and hence eliminated the problem of inexact copies), and a standard that could be relied upon never to change.

In 1961, by international agreement, a new standard of length was defined, based on the wavelength of a particular colour of light. Think, for example, of a sodium street lamp. No matter where the street lamp is situated—London or Glasgow, New York or Helsinki—it still emits the same yellow-coloured light. As you will see in Units 11–12, this is because the colour of the light is determined by the 'structure' of the sodium atoms—and sodium atoms are the same the world over. The scientific way of describing the colour of a light is in terms of its 'wavelength', which you will meet in Unit 10. In fact the substance chosen in connection with the length standard was not sodium, but krypton, whose atoms emit light of a characteristic orange-red colour.

So the length of the standard metre bar was carefully measured in terms of this 'wavelength of krypton light' and it was agreed that *exactly* 1 650 763.73 wavelengths would constitute one metre. This number was chosen so that the old definition of the metre (as the distance between the two inscribed marks on the platinum–iridium bar), and the new definition of the metre (as a particular number of wavelengths of krypton light) were kept in agreement. The advantage of the new definition was that it provided a standard of length far more precise than the metre bar. In addition, the krypton standard was readily available to laboratories all over the world, since krypton is present, albeit in small amounts, in the Earth's atmosphere.

The krypton wavelength is the sort of unit that can be called a *natural* unit since it depends on a particular natural property—namely, the fact that all atoms of a particular species are identical, and consequently always emit light of exactly the same colour.

Even then, that wasn't the end of the story. During the 1970s, it became clear that measurements of length made with the latest technology were limited not by the measuring techniques themselves, but rather by the definition of the metre in terms of the krypton wavelength. In other words, the new measuring techniques were potentially *more* precise than the standard of length. So, in 1983, it was agreed to adopt yet another definition of the

SECOND

KILOGRAM

metre. This one was different from all those that had gone before, because instead of being based on a *measurement*, the new standard was tied to a *defined* value. The property chosen this time was the speed of light in a vacuum.

Einstein, in his special theory of relativity, said that the speed of light in a vacuum (i.e. in empty space) is the maximum speed at which energy (or matter) can be transferred from one place to another: nothing can travel faster than light. Physicists are now convinced that the speed of light in a vacuum is a fundamental constant—always the same, everywhere in the Universe—and therefore perfect for use as a basic standard of measurement.

So nowadays, the speed of light in a vacuum is a *defined* quantity: in one second light travels *exactly* 299 792 458 metres. If you're wondering why scientists didn't take advantage of what seems a golden opportunity to define the speed of light as a nice round number, the explanation is quite simple. They wanted to keep the new definition completely consistent with the previous, krypton-wavelength standard. The slight penalty for that is having to live with an unwieldy value for the speed of light.

2.3 STANDARDS OF TIME

The concept of length is relatively straightforward. It is essentially a geometrical concept—a distance between two points in space. Length is easy to measure, too. We can make a metre rule; we can move the metre rule from place to place; we can use the metre rule today and tomorrow and the next day; and we can measure a particular length as often as we like.

Time is a more difficult quantity to measure. An interval of time can be used only once, and then it's gone—unless, that is, we can find some process that repeats with a regular and countable pattern. You met such a process in Unit 1: the cycle of day following night. Unfortunately, there are slight variations in the Earth's orbital speed during the course of a year. These variations, in turn, cause the interval between successive culminations of the Sun also to vary throughout the year. That's hardly satisfactory from the point of view of standardization—it means that a 'summer day' and a 'spring day' do not correspond to quite the same time interval. So we must look for a better standard. One possibility is to choose the *mean solar day* as the standard. The mean solar day is the *average* (taken over a year) of the time the Earth takes to spin once on its axis, relative to the Sun. The division of this mean solar day into 24 hours, and each hour into 60 minutes, and each minutes into 60 seconds, then gives us the basic unit of time: the **second**. It is the second that has been adopted as the SI unit of time.

This way of defining a unit of time in terms of the solar day has been adequate for the majority of everyday applications, but it has proved unsatisfactory for very high precision work. For, in addition to the variation in the solar day caused by variations in the Earth's orbital speed, there is also a cumulative slowing down of the Earth's spin (probably caused by tidal friction), the net effect of which is to cause our solar clock to lose 15 milliseconds every 1 000 years.

So in 1967, a *natural unit of time* was adopted. Like the natural unit of length, this natural unit of time was based upon the identical nature of all the atoms of a particular species—in this case, the atoms of caesium. Every atom vibrates at a characteristic frequency. The second is now defined as the time required for a caesium atom to vibrate exactly 9 192 631 770 times. The world's first caesium clock (Figure 3) was developed at the National Physical Laboratory, Teddington, in 1967. It kept time to an accuracy of better than 1 second in 10 000 years. Current technology, however, is doing even better than this. There are now clocks capable of providing a precision of 1 second in 3 million years!

FIGURE 3 The world's first caesium clock. It was not exactly designed for domestic use, and did not have a conventional 'clock face', but it did have the virtue of keeping time to better than 1 second in 10 000 years!

2.4 STANDARDS OF MASS

The concept of mass is quite a difficult one. Intuitively, we tend to regard mass as a measure of the amount of matter in an object. However, as we shall see in Unit 3, if one tries to work out a theory of the way in which objects move and interact using that kind of definition of mass, problems soon arise. In fact, a proper definition of mass requires an understanding of the physics of moving objects.

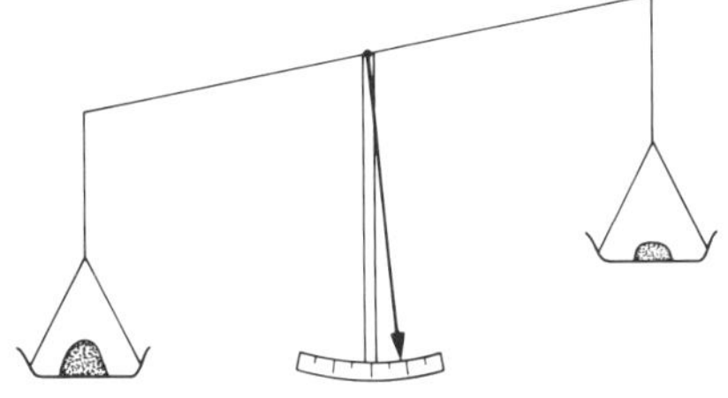

FIGURE 4 Schematic diagram of a beam balance used for comparing masses.

Fortunately for our purposes here, none of this creates a barrier to defining a standard of mass. What has been agreed is that the mass of one particular lump of matter will be called one **kilogram** and that this will define the SI unit of mass. (The internationally agreed standard lump is actually a cylinder of platinum–iridium kept in the International Bureau of Weights and Measures at Sèvres.) We can then compare this standard lump with any other lump of matter by using, for example, a beam balance (Figure 4). When the beam is level, we *define* the mass of the second lump to be the same as that of the standard. Secondary standards of mass made in this way have been distributed to standards laboratories throughout the world.

Of course, once a primary standard kilogram has been decided upon, it is possible to make a whole range of secondary standards—not just for one kilogram, but also for 0.5 kg, 2 kg, 5 kg, 10 kg, etc. The procedure is outlined in Figure 5.

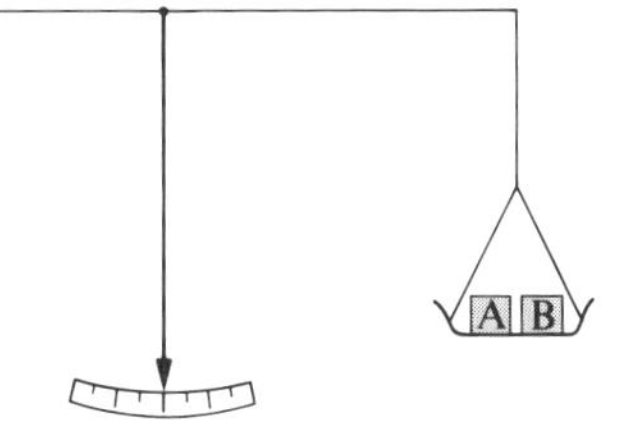

(a)

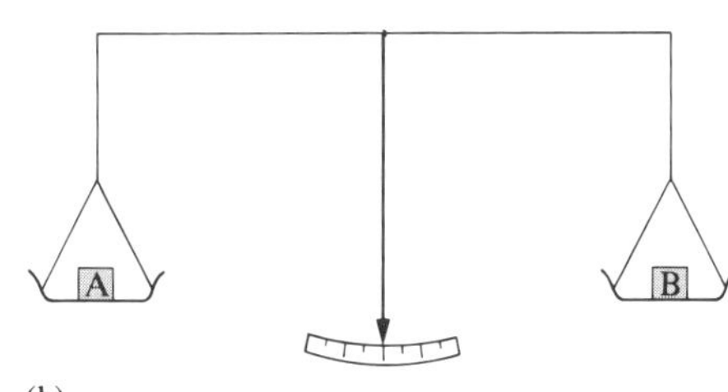

(b)

FIGURE 5 Generating a 0.5 kg standard, given a 1 kg standard. When the beam balances with *both* the arrangements of masses shown in (a) and (b), then: mass of A = mass of B = 0.5 kg

Naturally, it would be good if we could find an atomic standard of mass to supersede this operational standard. Such a standard does exist for the comparison of the masses of individual atoms, as you will find out in Units 11–12, but unfortunately we have not, as yet, discovered a way of scaling up this atomic mass standard with sufficient precision to allow us to use it for everyday mass comparisons. But doubtless we shall, one day!

POWERS-OF-TEN (SCIENTIFIC) NOTATION

ORDER OF MAGNITUDE

2.5 UNITS AND DIMENSIONS

2.5.1 THE POWERS-OF-TEN NOTATION

Since the choice of our basic units is essentially arbitrary, it would seem sensible to make the size of these units reflect the scale of things, or events, in our everyday experience. The metre, second and kilogram do this reasonably well. For example, a metre is about the size of a man's stride, a second approximately the time between consecutive human heartbeats, a kilogram the mass of a bag of sugar. However, when you consider that science is concerned with the whole range of natural phenomena—from events on a subatomic scale to developments on a galactic scale—you can see that we shall often meet quantities that are either very much smaller, or enormously larger than the size of these basic units. For instance, the distance to our nearest-neighbour star, Alpha Centauri, is about 40 400 000 000 000 000 metres. Or, at the other extreme, the time a radio signal takes to travel from London to Edinburgh is about 0.001 68 seconds. It is clearly inconvenient to have to write these quantities in this way, so scientists use a neat kind of shorthand—**powers-of-ten notation** (often called **scientific notation**).

The basic idea behind this notation is that several tens multiplied together generate very large numbers. In the powers-of-ten notation these numbers are represented by placing a superscript after the number 10; the superscript indicates the number of tens that have to be multiplied together to get the number. That is:

10^1 means $10 = 10$

10^2 means $10 \times 10 = 100$

10^3 means $10 \times 10 \times 10 = 1\,000$

10^4 means $10 \times 10 \times 10 \times 10 = 10\,000$

10^5 means $10 \times 10 \times 10 \times 10 \times 10 = 100\,000$

This superscript is called the *power* to which ten is raised.

We can cope with very small numbers in much the same way by taking the *reciprocal* of several tens multiplied together. The fact that a reciprocal has been taken is indicated by a negative sign in front of the superscript, that is:

$$0.1 = \frac{1}{10} = \frac{1}{10^1} \text{ is written as } 10^{-1}$$

$$0.01 = \frac{1}{100} = \frac{1}{10^2} \text{ is written as } 10^{-2}$$

$$0.001 = \frac{1}{1\,000} = \frac{1}{10^3} \text{ is written as } 10^{-3}$$

and so on.

TABLE 1 Numbers expressed as powers of ten

$1\,000 = 10^3$
$100 = 10^2$
$10 = 10^1$
$1 = \ldots$
$0.1 = 10^{-1}$
$0.01 = 10^{-2}$
$0.001 = 10^{-3}$

With this notation, the number 1 can also be accommodated. Look at Table 1. In this sequence of numbers, it is obviously logical to define the number 1 by

$$1 = 10^0$$

Indeed, *any* number raised to the power zero is defined to be equal to 1.

The powers-of-ten notation can also be used when the number in question is not an exact multiple of ten. For instance, we can write:

$$4\,000 = 4 \times 1\,000 = 4 \times 10^3$$

Similarly:

$$0.06 = 6 \times (1/100) = 6 \times 10^{-2}$$

Consequently, we can now write the distance to Alpha Centauri as 4.04×10^{16} m, and the time for the radio signal to travel from London to Edinburgh as 1.68×10^{-3} s. These numbers are much more easily assimilated in this form than in the long form in which they incorporate a large number of zeros.

ITQ 1 Use scientific notation to express the number of seconds in one solar day (i.e. 24 hours).

Because scientists frequently use this kind of notation, they have devised a system of names and abbreviations for some of the powers of ten to be applied as prefixes to the basic units of measurement (Table 2). Thus one thousand metres can be said as one kilometre and written as 1 km. One thousandth of a second is said as one millisecond and written as 1 ms.

TABLE 2 Prefixes to units of measurement

Prefix*	Symbol	Equivalent of prefix in powers-of-ten notation
tera	T	10^{12}
giga	G	10^{9}
mega	M	10^{6}
kilo	k	10^{3}
centi	c	10^{-2}
milli	m	10^{-3}
micro	μ	10^{-6}
nano	n	10^{-9}
pico	p	10^{-12}
femto	f	10^{-15}

* Note that when a prefix is placed in front of a unit, in effect it produces a new unit. Consequently, km^2 (for instance) should be read as $(\text{kilometres})^2$ and not as kilo $\times$ $(\text{metres})^2$; thus $1\ km^2 = (10^3\ m)^2 = 10^6$ square metres.

As you can see from Table 2, the prefixes generally change in steps of 10^3 (i.e. 1 000). This is the preferred SI convention. However, the intermediate prefix *centi* is such common usage that it has also been included in the Table. You will need the information in Table 2 many times during the Course, so for ease of reference it is reprinted on the back cover of the Glossary.

□ Light takes approximately 3.34×10^{-7} seconds to travel 100 metres. What is this time in nanoseconds?

■ Because 1 nanosecond $= 10^{-9}$ seconds,

it follows that

$$1 \text{ second} = 10^9 \text{ nanoseconds}$$

Hence

$$3.34 \times 10^{-7} \text{ seconds} = (3.34 \times 10^{-7}) \times 10^9 \text{ nanoseconds}$$
$$= 334 \text{ ns}$$

2.5.2 ORDERS OF MAGNITUDE

Scientists sometimes call a power of 10 an **order of magnitude**. So, for instance, we could say that £1 is an order of magnitude more valuable than a 10p piece, or *two* orders of magnitude more valuable than a 1p piece. However, it is more common for this 'order of magnitude' expression to be used in an *approximate* sense. For example, since the distance from London to Aberdeen is about 490 miles and the distance from London to Milton Keynes about 55 miles, we might say that Aberdeen is an order of magnitude further away from London than is Milton Keynes. For many purposes this kind of statement would be quite accurate enough.

Similarly, in science, it is often very useful to be able to get just a rough idea of the size of some quantity, without having to do an exact calculation or to carry out a very careful experiment. In fact, the idea of quoting quantities to 'within an order of magnitude' is so useful that a special symbol has been devised to represent this kind of relationship. We write: the distance to Alpha Centauri $\sim 10^{16}$ m. The symbol $\sim$ means 'is of the same order of magnitude as'. This statement tells us that the distance involved is 10^{16} m *to within a factor of ten.*

DIMENSIONS

□ How could we calculate, to within an order of magnitude, the number of seconds in the lifespan of a typical adult in Britain?

■ We can do this quite quickly by making approximations to get numbers that are easy to handle. If we were to assume a typical lifespan of 70 years, the full calculation would be

70 years $\times$ $365\frac{1}{4}$ days per year $\times$ 24 hours per day $\times$ 60 minutes per hour $\times$ 60 seconds per minute

If you were to work this out on your calculator you would find it came to 2 209 032 000 seconds. Expressed as an order of magnitude (i.e. to the nearest integral power of ten) this is 10^9 s.

However, there is no point in making such an accurate calculation if all we want is an order of magnitude estimate. It is much simpler to approximate, indeed very crudely, so as to make the numbers as easy as possible:

$$70 \text{ years} \approx 50 \times 400 \times 25 \times 60 \times 60 \text{ seconds} = 1\,800\,000\,000\,\text{s}$$

As an order of magnitude, this too is 10^9 s.

The ability to make this sort of estimate is a very useful one for a scientist. In fact, it is a good habit to check the result of every calculation by doing a rough approximation, just to ensure that the answer you get is sensible.

TABLE 3 The meaning of some mathematical symbols

Symbol	Meaning
$=$	is equal to
$\approx$	is approximately equal to
$\sim$	is of the order of magnitude of
$>$	is greater than
$<$	is less than
$\gtrsim$	is greater than or roughly equal to
$\lesssim$	is less than or roughly equal to
$\geqslant$	is greater than or equal to
$\leqslant$	is less than or equal to

ITQ 2 You calculated in ITQ 1 that there are 8.64×10^4 seconds in one day. How may seconds are there, *to within an order of magnitude*, in 1 week?

The mathematical symbols that you are likely to come across in this Course are listed in Table 3. You will have already met the familiar 'is equal to' sign. The $\approx$ sign, meaning 'is approximately equal to' is not quite as loose as the $\sim$ sign. It implies the *rounding off* of a quantity, rather than a possible factor-of-ten uncertainty. For example, one could write

$$26.7 \approx 27 \quad \text{or} \quad 267 \approx 270$$

but $267 \sim 300$

2.5.3 DIMENSIONS

Before leaving this subject of units and standards, there is one final point that is worth mentioning: *every measured quantity must have units associated with it.* Making a measurement implies comparing the quantity being measured with some agreed standard, so the measured quantity must take on the same units as the standard. This is a point you should take to heart: wherever you write down the numerical value of a measurement, it is essential always to write down the units as well.

Note the importance of the word 'measured' in the previous paragraph. Directly measurable quantities have units, but their ratios and multiples do not necessarily do so. For instance the ratio of two lengths is a pure number:

$$\text{e.g.} \quad \frac{50\,\text{m}}{25\,\text{m}} = 2 \quad \text{and} \quad \frac{50\,\text{km}}{25\,\text{m}} = \frac{50 \times 1\,000\,\text{m}}{25\,\text{m}} = 2\,000$$

These apparently trivial examples are important because they show that units can be multiplied and divided by one another. We make use of this fact whenever we read the speedometer in a car. If you've driven steadily along the motorway and noted that the time taken between successive mile markers was 1 minute, you'd instantly calculate your speed as:

$$\frac{1 \text{ mile}}{1 \text{ minute}} = \frac{60 \text{ miles}}{60 \text{ minutes}} = \frac{60 \text{ miles}}{1 \text{ hour}} = 60\,\text{m.p.h.}$$

Thus whenever you divide one quantity by another, you divide not only the numbers but also their respective units. So, 10 kilometres in 5 hours is 2 km/hour (i.e. 2 kilometres per hour). Similarly, whenever you multiply two quantities together, you must also multiply their respective units. So, 5 metres × 12 metres is 60 metres^2.

□ What is $3\,\text{mm} \times 2\,\text{m}$, in units of m^2?

■ $$3\,\text{mm} = 0.003\,\text{m}$$

So

$$(3\,\text{mm}) \times (2\,\text{m}) = (0.003\,\text{m}) \times (2\,\text{m}) = 0.006\,\text{m}^2$$

Alternatively, we can reason that

$$3\,\text{mm} = 3 \times 10^{-3}\,\text{m}$$

So

$$(3\,\text{mm}) \times (2\,\text{m}) = (3 \times 10^{-3}\,\text{m}) \times (2\,\text{m}) = 6 \times 10^{-3}\,\text{m}^2$$

The importance of all this is that it provides us with a very powerful tool for checking whether an equation relating two physical quantities is likely to be correct—the technique called *dimensional analysis*. This technique is based on a recognition of the fact that we can only equate two quantities if they are both lengths, or both speeds, or both times or whatever—we can only equate like with like. The scientist's way of saying this is that the quantities must have the same **dimensions**.

Why the special word 'dimensions' instead of just 'units'? The reason is quite simple: the use of dimensions allows us to equate quantities expressed in units that differ only by a conversion factor. For example, feet, miles and metres are all different units, yet they have a common dimension—length. Similarly, hours and seconds, though different units, both have the dimension of time. To a good approximation:

$$50\,\text{m.p.h.} = 22.35\,\text{m/s}$$

This equation, which balances miles per hour against metres per second, is perfectly valid (provided, of course, that the appropriate conversion factor has been incorporated into the equation). For although the units do not match exactly, the dimensions do: both sides of the equation have dimensions of length divided by time.

SUMMARY OF SECTION 2

In this Section, you have been introduced to some of the scientific and mathematical skills and techniques that you will need later in this Unit, as well as in subsequent Units. In particular, you have seen how to:

1 handle reciprocals, multiples and fractions of quantities;

2 combine quantities by multiplying or dividing one by another;

3 express a number using the powers-of-ten notation;

4 convert the units of a physical quantity to other equivalent units using (where possible) standard prefixes and powers of ten;

5 use the order-of-magnitude symbol appropriately;

6 identify the basic dimensions of a quantity, and so check the validity of equations involving that quantity.

You have also seen that any measured quantity must have units associated with it, and have been introduced to the measurement standards and SI units for length, time and mass (the metre, the second and the kilogram, respectively).

3 THE EARTH, THE SUN AND THE MOON

3.1 INTRODUCTION

Let us now turn from the theory of measurement to its practice. In Unit 1 you saw how our present-day model of the Solar System was in qualitative agreement with observation. But, of course, the model of the Solar System was not developed purely on the basis of qualitative observations. Historically, qualitative observations were often mixed in with measurements—measurements that were sometimes accurate and sometimes wildly inaccurate (when viewed with hindsight). Sometimes the inaccurate measurements actually held back the development of the model. Why were some of the measurements so inaccurate? As you might guess, the answer to this question is not simple. In part, the inaccuracies can be attributed to poor tools and observational aids. A lot of the difficulty, however, also lay in the indirect way in which the measurements were made, or in the rather cavalier manner in which the data were extended beyond the range of the actual measurements. Sometimes dubious assumptions were used in calculating the required quantity from the measurements.

This, however, is not to devalue the work of our ancestors! Indeed, the tricks they used then are, in many ways, very similar to the tricks we use ourselves today. Whenever the measurement to be made lies outside the scope of current observational techniques, we have to rely on our initiative to devise and interpret the results of indirect methods of measurement. Initiative was something the early 'measurers of the Solar System' had in abundance. So perhaps we can learn something by looking at the sort of measurements they made, and the sort of reasoning they employed.

3.2 THE SIZE OF THE EARTH

3.2.1 THE SCIENTIFIC SCHOOL AT ALEXANDRIA

The city of Alexandria was founded at the mouth of the River Nile by Alexander the Great, during his conquest of Asia Minor, Egypt and Persia. It became an important centre of learning during the fourth, third and second centuries BC, and the Museum of Alexandria (which functioned like an academy and university) attracted some of the foremost Greek scholars of the time. In about 300 BC, a school of astronomy was established there, a school that was to bring a new attitude to the science. For these astronomers sought to *quantify* the sizes and distances involved in the Solar System and so to turn astronomy into a 'real' science. Conjecture was all very well, they argued, but it must be based on measurement.

The obvious starting point was to try to find the size of the Earth. How big was the world they lived on? One of the earliest estimates was provided in about 235 BC by the Greek astronomer Eratosthenes. His measuring technique was, of necessity, indirect. After all, very little of the world was known to the Greeks in 235 BC. And, of course, Eratosthenes did not have the advantage of our modern technological aids such as radio communications and space flight. Yet, starting from the assumption that the Earth was spherical,* he managed to get a value for the circumference (and thus also for the radius) of the Earth that compares very favourably with the currently accepted value. (He was probably within 5% of the modern value.)

* The idea of a spherical Earth was well established in Greek culture. Aristotle had argued that, on the grounds of symmetry alone, the Earth *must* be a sphere. But there was also the experimental evidence provided by (a) the always circular shape of the Earth's shadow thrown onto the Moon during a lunar eclipse, and (b) the change in the position of the stars as an observer travelled northwards or southwards (see Unit 1). It was not until the Middle Ages that the 'flat Earth' idea again became popular.

3.2.2 HOW ERATOSTHENES DETERMINED THE RADIUS OF THE EARTH

Eratosthenes approached the problem by assuming the Sun was so far away that all 'sunbeams' reaching the Earth were, in effect, parallel. (We shall examine this assumption in detail in the TV programme associated with this Unit.) He then simultaneously compared the direction of the vertical at two different locations on the Earth's surface, with the direction of the parallel beams of sunlight at that instant (Figure 6). But how did Eratosthenes make *simultaneous* measurements at two different places on the Earth's surface? If the Greeks had had reliable clocks that could be synchronized and then transported, he could have asked an assistant to make the measurement at one of the locations at some previously agreed instant of time (as shown by one of the clocks), while he made the measurement at the other location at the same instant of time (as shown by the second clock). But the Greeks had no such clocks. So Eratosthenes solved the problem of synchronization by carrying out his measurements at noon (i.e. the time when the Sun was highest in the sky) at two places lying on what we should now call the same line of longitude.

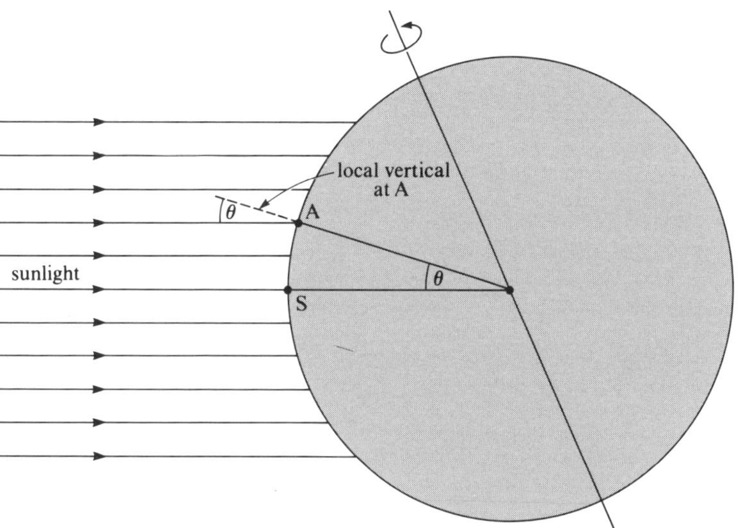

FIGURE 6 The parallel beams of sunlight provide a *reference* direction for all points facing the Sun. The direction of the local vertical (i.e. the direction in which a plumb-line would hang at that locality) is, by definition, the direction determined by a line passing through that point on the Earth's surface, and the centre of the Earth. In other words, the local vertical lies in the same direction as the Earth's radius at that location (plumb-lines point towards the centre of a spherical Earth).

□ Why did he choose two locations lying on the same line of longitude?

■ A line of longitude is a line drawn around the Earth in a north–south direction and passing through the two Poles. Consequently, all points on this line experience noon (i.e. that time of day when the Earth's rotation brings the Sun to its culmination point) at the same instant in time.

The two locations that Eratosthenes chose were Alexandria (where he worked) and Syene—now called Aswan—almost exactly 500 miles due south of Alexandria (A and S in Figure 6). It is easy for us to quote this distance nowadays, but Eratosthenes must have found the measurement very difficult to make. We are not sure exactly how he solved the problem—the various documented accounts are in disagreement on this point. Whatever technique he did actually use, it was fortunate that the terrain between Alexandria and Syene was sufficiently flat to allow the measurement to be made at all.

So what observations did Eratosthenes make at Alexandria and Syene, and how did his observations enable him to estimate the size of the Earth? Well, the observation at Syene was a very simple one. He knew from records kept at Syene that, at exactly noon on Midsummer's Day, sunlight falling on a

DEFINITION OF π

very deep well there reached the water surface and was reflected straight back up the well again. (What the records said was that the water in this particular well was only visible at noon on Midsummer's Day. We would now say that Syene lies on the Tropic of Cancer.)

What do you think Eratosthenes deduced from this fact?

Eratosthenes reasoned that the direction of the sunlight and the direction of the local vertical at Syene coincided at that particular instant (Figure 6). That is, at noon on Midsummer's Day, the Sun at Syene was exactly 'vertically' overhead. Hence the angle between the direction of the Sun's rays and the direction of the local vertical was zero degrees. (One degree is defined to be 1/360 of a rotation through a complete circle. In scientific texts one degree is usually written as 1°.)

Consequently, if Eratosthenes also measured the angle between the direction of the Sun's rays and the direction of the local vertical at *Alexandria* at noon on Midsummer's Day, he would, in practice, be measuring the angle between the Earth's radius to Syene and the Earth's radius to Alexandria. These two (equivalent) angles are both labelled θ in Figure 6. (θ, the Greek letter pronounced 'theta', is frequently used to denote angles.)

Then, however, Eratosthenes was faced with another problem. It is easy to say: 'measure the angle between the Sun's rays and the local vertical'. Yet how was he to do this in practice?

His trick was to use the fact that unless the Sun is directly overhead, it casts shadows. So, by placing a vertical pole, whose height he knew, in the ground at Alexandria, and by measuring the length of the shadow cast by this pole at noon on Midsummer's Day, Eratosthenes was able to deduce that the angle between the Sun's rays and the vertical at Alexandria was 7.5° (Figure 7). Hence the angle between the two Earth radii—to Alexandria and to Syene—is also 7.5°. Alexandria and Syene are separated by 500 miles. So, if 7.5° between two Earth radii correspond to 500 miles around the Earth's circumference (Figure 8), then 1° between radii would correspond to (500/7.5) miles around the circumference.

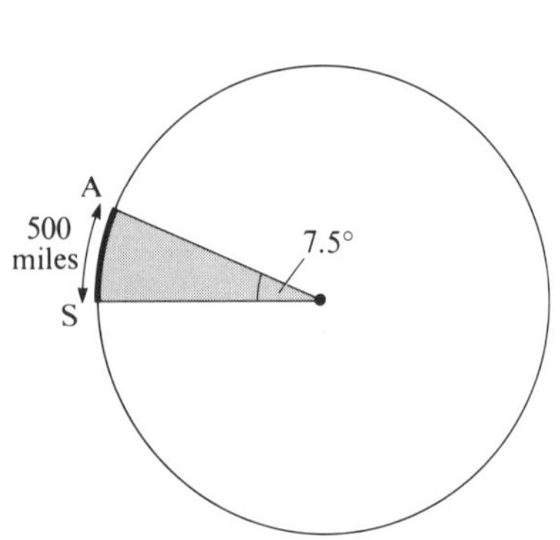

FIGURE 8 A is Alexandria and S is Syene. An angular separation of 7.5° between Earth radii corresponds to a distance of 500 miles around the circumference of the Earth. (Note that, to increase clarity, the size of the angle has been exaggerated in this diagram.)

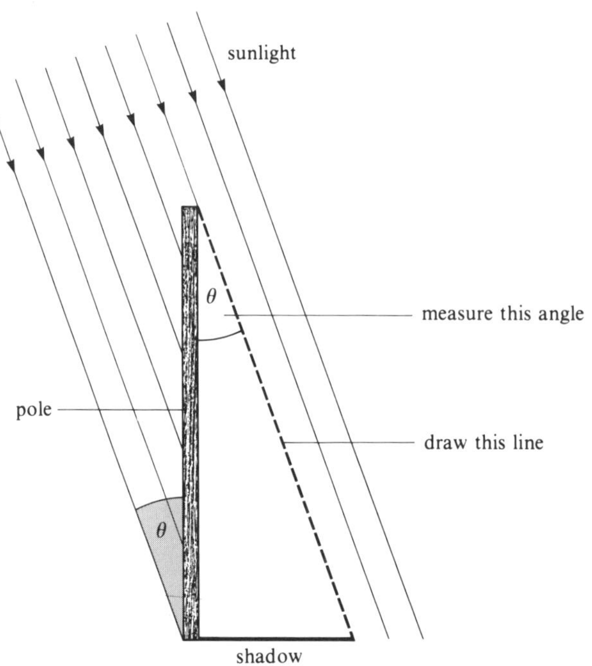

FIGURE 7 If we know the height of the pole and also the length of the shadow cast by it, we can draw a scale diagram (like the one shown here) to enable us to measure the angle θ between the vertical (the direction of the pole) and the direction of the Sun's rays.

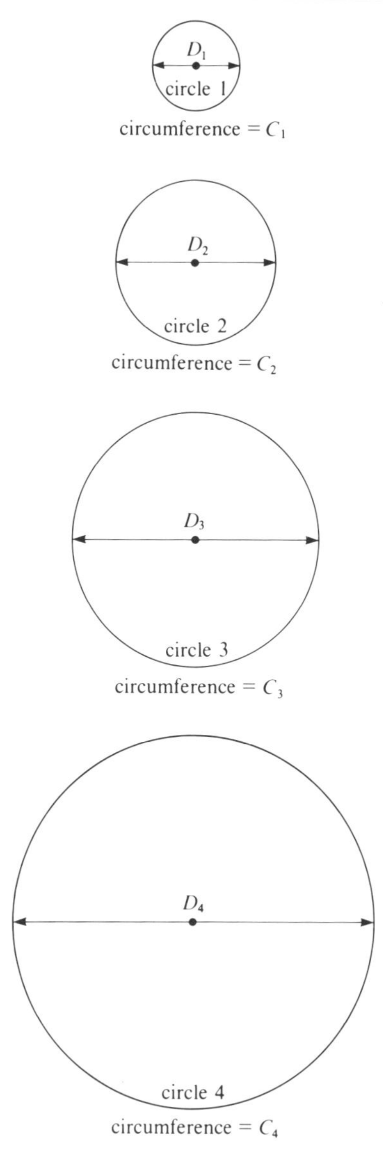

FIGURE 9 The **definition of π**. You will also find π discussed in *MAFS 4*.

ITQ 3 What distance around the circumference, therefore, would 360° between Earth radii correspond to?

So, in this way, Eratosthenes was able to determine the approximate circumference of the Earth. Finding the radius was then a relatively easy step. The Greeks had long known that, for *any* circle, no matter how large or how small, the ratio between the circumference of that circle and its diameter is always the same (Figure 9). This fixed ratio is now denoted by the Greek letter π (pi).

$$\frac{\text{circumference}}{\text{diameter}} = \pi \qquad (1)$$

This can be rewritten to give an expression for the circumference, by multiplying both sides of the equation by the diameter of the circle:

$$\frac{\text{circumference}}{\text{diameter}} \times \text{diameter} = \pi \times \text{diameter}$$

or

$$\text{circumference} = \pi \times \text{diameter} \qquad (2)$$

It becomes very tedious having to write out the words circumference and diameter all the time. So it is common practice to represent these quantities by letters. If the circumference is denoted by C and the diameter by D, then Equation 2 becomes

$$C = \pi \times D$$

or

$$C = \pi D \qquad (3)$$

(Notice that the multiplication sign between π and D is not necessary. If you see two symbols next to each other like this, the multiplication sign is implied.)

The final step is to realize that the diameter of a circle is just twice the radius. So if the radius is represented by the symbol R, then:

$$D = 2R \qquad (4)$$

and Equation 3 can be written as

$$C = 2\pi R \qquad (5)$$

(where we have simply replaced D by $2R$—the order in which quantities are multiplied together is immaterial).

You probably recognize Equation 5. Anyway, the important thing to notice is that if the circumference of the Earth is known, the radius of the Earth can be calculated.

ITQ 4 In ITQ 3 you calculated Eratosthenes' value for the circumference of the Earth. If π is taken to be 3.14, what (according to Eratosthenes) is the *radius* of the Earth?

ITQ 5 Given that 1 mile is approximately equal to 1.61 km, convert your answer to ITQ 4 into kilometres.

3.2.3 AN ALTERNATIVE PERSPECTIVE

There is an alternative way of analysing the problem described in Section 3.2.2, which is well worth taking a look at. To understand it, you need to learn some new mathematical techniques, but now that you have already seen the basic principles of Eratosthenes' method, and have worked out approximate values for the circumference and radius of the Earth, you are in a good position to follow through this alternative approach. Furthermore, the mathematics that you will learn en route will be needed later in this Unit.

RADIAN

SMALL-ANGLE APPROXIMATION

The starting point for this alternative approach is the definition of the units of angular measurement. In the preceding Section, all angles were measured in *degrees*, where one degree (1°) was defined to be 1/360 of a complete rotation. However, it is often more convenient in science to use a different unit of angular measurement—the *radian*. A radian is defined in the following way.

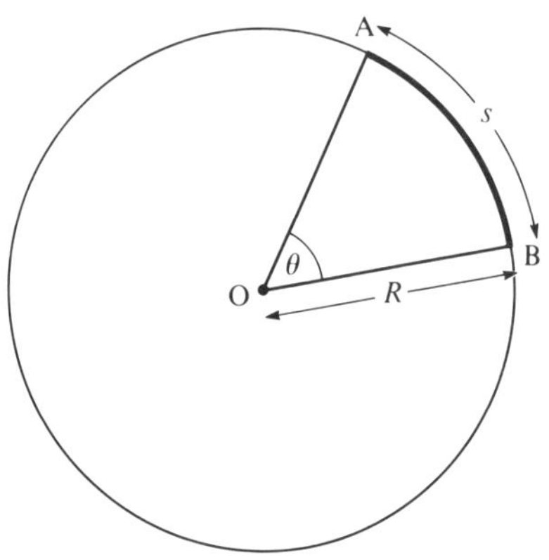

FIGURE 10 The angle θ encloses an arc of length s. Conversely, the arc s *subtends* an angle θ at the centre of the circle. The angle subtended when $s = R$ is defined to be 1 radian.

For a general angle of θ radians, $s = R\theta$ (see text).

Figure 10 shows a circle of radius R with two radii separated by an angle θ. That part of the circumference of the circle enclosed by the two radii is known as an *arc* of the circle. (This arc is said to *subtend* an angle θ at the centre of the circle.) If the angle θ is adjusted until the length s of this arc (as measured *along* the circumference) is exactly equal to R, then we say that θ is defined to have a value of one **radian**.

At first sight this seems to be a very complicated way of defining a unit of angle, but there is method in the madness! First, if an angle of one radian means that the arc length is *equal* to the radius, then an angle of two radians means that the arc length is equal to *twice* the radius, and an angle of three radians means that the arc length is equal to *three times* the radius, and so on. In fact, for any circle of fixed radius R, an increase in θ gives a proportionate increase in arc length. The equation describing this relationship is:

$$s = R\theta \tag{6}$$

where θ is measured in radians.

Equation 6 is an equation you will come across quite frequently—you should memorize it. It says *arc length equals radius multiplied by subtended angle (in radians)*. That is not all, however. Recall that the circumference of a circle is equal to twice the radius of that circle multiplied by π; that is:

$$C = 2\pi R \qquad (5)^*$$

(The asterisk to the right of the equation number denotes an equation that has already appeared earlier in the text. This convention is used throughout the Course.)

A circumference, however, is simply an arc *that goes all the way round the circle*. So Equation 5 is a special case of Equation 6, in which the arc length, and hence the angle, corresponds to a complete circle. By comparing the two Equations:

$$s = R\theta \qquad (6)^*$$

and $$C = 2\pi R \qquad (5)^*$$

$$= R \times 2\pi$$

we must conclude that the angle corresponding to a complete circle is 2π *radians*. Consequently:

$$\boxed{360^\circ = 2\pi \text{ radians}} \tag{7}$$

ITQ 6 A right angle is defined to be 90°. Express a right angle in radians.

ITQ 7 Use Equation 7 to show that 1 radian $\approx$ 57.3°.

ITQ 8 Use Equation 6 to deduce the *dimensions* of angle. Refer back to Section 2 if you've forgotten what is meant by the term 'dimensions'.

How can all this now be applied to Eratosthenes' data? Look at Figure 11. You have already seen that the angle between the direction of the Sun's rays and the vertical direction of the pole at Alexandria was the same as the angle between the two Earth radii to Alexandria and Syene respectively. So if this was 7.5° and the distance from Alexandria to Syene was 500 miles . . .

FIGURE 11 A is Alexandria and S is Syene. For clarity, the height and shadow of the pole at Alexandria and the angle θ have all been exaggerated in the diagram.

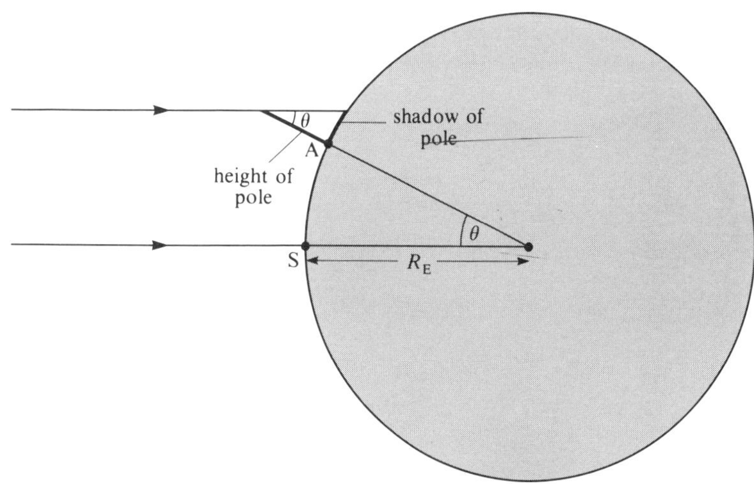

ITQ 9 Using Equation 6 this time, calculate R_E, the radius of the Earth, in miles. (The subscript E is used to remind you that the radius in question is the radius of the Earth. This type of subscript notation is very common in science.)

The only thing that is a bit dubious about this last calculation is the value of the angle θ. How was θ actually measured? It was suggested earlier that one way would be to measure the height of the vertical pole, and the length of the shadow cast by the pole, and then to draw a scaled-down diagram. The angle θ would then have to be measured with a protractor. But this diagram is not really necessary. If we don't mind making a slight approximation in our calculations, we can find the value of θ *directly* from the measurements of shadow length and pole height. We can say that:

$$\text{shadow length} \approx \text{pole height} \times \theta \text{ (in radians)} \qquad (8)$$

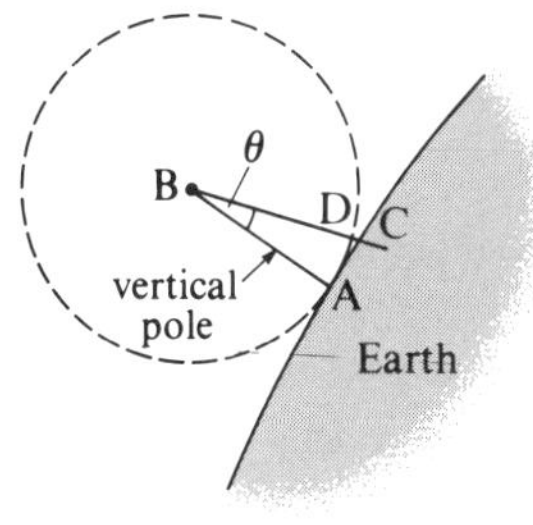

(a)

Why? Look at Figure 12a. For the circle centred on the tip of the pole (i.e. B), using Equation 6 we can write:

$$\text{arc AD} = \text{radius BA} \times \theta$$

We are not, however, particularly interested in the length of the arc AD—we are much more concerned with the length of the shadow (AC) along the surface of the Earth. (In this notation 'line BA' means the line whose ends are defined by the points B and A.)

Look at Figure 12b, which is an enlargement of Figure 12a. On this scale you can see that the shadow AC is a straight line.* Furthermore, the length of the arc AD is almost indistinguishable from the shadow length AC. In fact, in order to make the difference between AC and AD visible, the angle θ in this Figure has been made greater than 15°, whereas Eratosthenes' angle was about 7.5°!

So, in summary, if θ is small, the arc AD can be treated as being approximately equal to the length of the shadow along the surface of the Earth. That is why Equation 8 is a valid approximation. In fact, when θ is small (less than about 15° or 0.26 radians), the curved arc length in Equation 6 can be approximated by a straight line. The smaller the angle, the better the approximation. (An angle of 15° leads to an inexactness of about 1%.) This approximation is sometimes called the **small-angle approximation**.

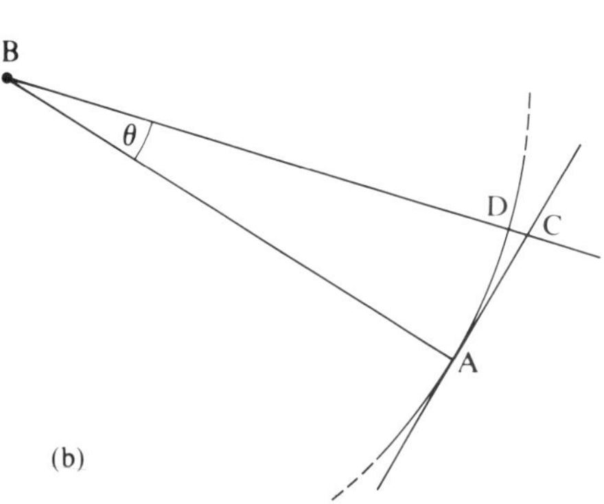

(b)

FIGURE 12 If the angle θ is small enough, the length of the arc AD is approximately the same as the length of the straight line AC.

If we now rearrange Equation 8 to give an expression for θ, we find that, for small values of the angle,

$$\theta \approx \frac{\text{shadow length}}{\text{pole height}} \qquad (9)$$

that is, the value of the angle (in radians) can be determined by calculating the ratio of shadow length to pole height.

* The curvature of the Earth will have a negligible effect for these small distances.

UPPER AND LOWER LIMITS

UNCERTAINTIES IN A MEASUREMENT

ITQ 10 You have already seen that Eratosthenes found θ to be 7.5°. Deduce what length of shadow would have been cast at Alexandria by a pole of height 100 cm.

At this stage it is worth pausing to summarize the main points of Section 3.2:

1 Two units of angular measure are frequently used by scientists: the *degree*, which is 1/360 of a complete rotation and the *radian*, which is $1/2\pi$ of a complete rotation. They are related by the equation:

$$360^\circ = 2\pi \text{ radians.}$$

2 The length s of an arc of a circle of radius R is related to the angle θ subtended by that arc at the centre of the circle, by the equation:

$$s = R\theta$$

where θ must be measured in radians.

3 When θ is small the curved arc length s can be taken as a straight line (the small-angle approximation).

3.3 THE RADIUS OF THE MOON

The Earth's nearest neighbour is the Moon. How does the size of the Moon compare with the size of the Earth? How *can* the size of the Moon be compared with the size of the Earth? The early astronomers knew that when the Earth passed between the Sun and the Moon a shadow of the Earth was thrown onto the Moon (Figure 13a). Figure 13b shows 20th century photographs of such a lunar eclipse. *If it is assumed that the shadow of the Earth on the Moon is the same size as the Earth itself*, then the ratio of the size of the Moon to the size of the Earth can be estimated from Figure 13b by completing the circle of the Earth's shadow, and finding the ratio of the radius of this circle to the radius of the circle of the Moon. You should now think about how you might use a pair of compasses and a pencil* to complete the circle of the Earth's shadow.

You will probably have some difficulty deciding exactly what size of circle best fits the arc of the shadow. The best way out of this difficulty is to draw not just one circle, but *two*. The first circle should be the *biggest* circle that you feel could be fitted to the arc; the second circle should be the *smallest* circle that could fit the arc. By doing this you will have estimated, not the *exact* size of the Earth's shadow, but the **upper and lower limits** to its possible size. You have found the size of the shadow to within certain tolerances. You will find that this is something you often have to do in science, namely to estimate the *range* of possible values that your measurement could cover. No measurement is ever exact—there will always be some **uncertainties** associated with it. So if you can say what the limits of these uncertainties are, other people do at least know what sort of credibility to give your measurement.

What values do you get for the upper and lower limits of the radius of the Earth's shadow? (Make the measurements in centimetres.)

The radius of the Earth's shadow in the photograph is:

less than.................... cm (upper limit)

more than.................. cm (lower limit)

* If you are at all uncertain about how to use a pair of compasses to draw a circle, you should refer to *Into Science*, Module 9.

FIGURE 13 (a) The Earth, lying between the Sun and the Moon, casts a shadow on the Moon's surface. (As the TV programme shows, the way the Earth's shadow has been drawn here is not exactly correct, but this representation is adequate for the present argument.)

(b) Photographs showing three partial phases of the lunar eclipse of 2 May 1920. Notice that the less of the Earth's shadow you have, the more difficult it is to estimate R_E. On the other hand, the more of the Earth's shadow you can see, the more difficult it is to measure R_M.

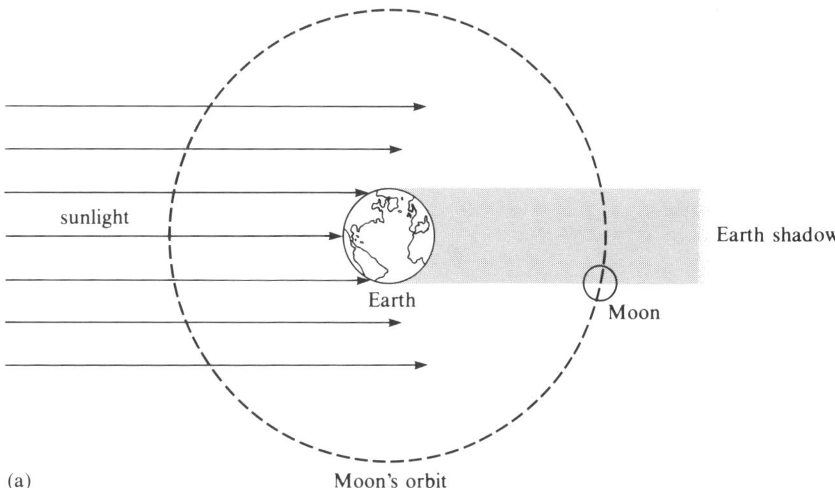

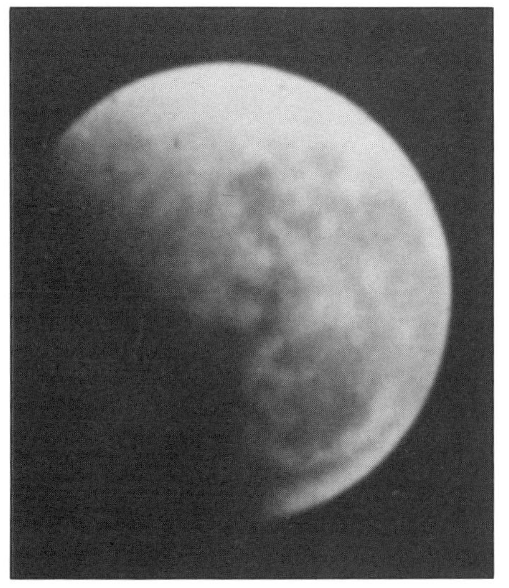

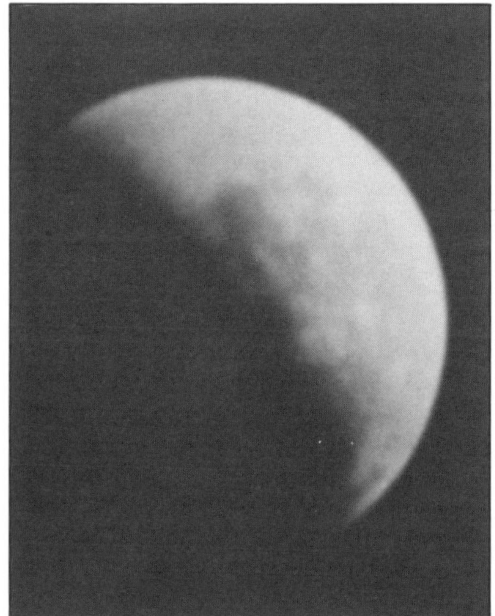

(b)

ITQ 11 What, therefore, would you say is the *best estimate* you can give for the radius of the Earth's shadow in the photograph?

You will need to use this value again, so when you have completed ITQ 11 (and checked your answer against the one at the back), write your estimate in the space below.

Radius of Earth's shadow in the photograph = ... ± .... cm

The next measurement you need is that of the radius of the Moon (as shown in the photograph). Again make this measurement in centimetres:

Radius of the Moon in the photograph = ... cm

☐ What do you estimate are the upper and lower limits for this measurement?

■ In this case, you probably felt that you could measure the radius of the Moon (in the photograph) to within about ±0.1 cm. This is a much smaller uncertainty than that involved in the measurement of the radius of the Earth.

So (given the assumption stated on p. 20) you can now say what the ratio is between the radius of the Earth and the radius of the Moon.

The radius of the Earth is .... times bigger than the radius of the Moon; that is,

radius of the Earth = ... Moon radii

ANGULAR SIZE

ITQ 12 Denoting the Earth's radius by R_E and the Moon's radius by R_M, what are the limits of uncertainty involved in your value for the ratio R_E/R_M?

When you have worked through ITQ 12, and checked that your answer is reasonable, write your final result in the space below:

$R_E = (\ldots\ldots\ldots\ldots\ldots \pm \ldots\ldots\ldots\ldots\ldots)R_M$

It is an easy matter now to convert the radius of the Moon (expressed as a fraction of the radius of the Earth) into a measurement in kilometres.

ITQ 13 Find the approximate radius of the Moon in kilometres, assuming that the radius of the Earth is about 6 200 km. (Eratosthenes' value of 6 150 km (ITQ 5) was probably up to 5% out. So, as we only want approximate values here, it is perfectly all right to work with the 'rounded-up' figure of 6 200 km.)

Again, write your final answer in the space below; don't forget to estimate the uncertainty in your value.

$R_M = \ldots\ldots\ldots\ldots\ldots \pm \ldots\ldots\ldots\ldots\ldots$ km

One word of caution about this calculation of R_M: *throughout Section 3* you make the assumption that the shadow of the Earth at the Moon is the same size as the Earth itself. As you will see in the TV programme, this is an *unjustified assumption* that leads to a considerable error in the value obtained for R_M. However, the programme also shows how to correct this error, so be prepared to adjust your value of R_M after viewing. (The essential details of this correction are summarized in the TV Notes, at the end of the text.)

3.4 THE DISTANCE TO THE MOON

3.4.1 ECLIPSING THE MOON

Now we know the approximate radii of both the Earth and the Moon. The next measurement we want to make is the *distance between* the Earth and the Moon. Apart from the Sun, the Moon appears as the largest body in the sky. But just because it *appears* to be the largest body, it does not necessarily follow that it *is* the largest body. Indeed, you probably already have a suspicion that the Moon appears larger than the stars because it is nearer than the stars! Apparent size must obviously depend not only on the real size of the object, but also on the distance of a object from you, the observer. The important point to realize is that the apparent size of the Moon is determined by the angle it subtends at your eye.

What actually determines the angle? Look at Figure 14. The **angular size** of the Moon in this diagram (i.e. θ_M) is clearly dependent both on the diameter of the Moon (D_M) and on the distance of the Moon from the observer on Earth (i.e. L_M). Once again, we can use the relation: arc length equals radius multiplied by subtended angle in radians. In this case the centre of the circle is point O in Figure 14, so the radius is L_M, the arc length is D_M and the subtended angle is θ_M. Admittedly, this is an approximation; but if θ_M is small, the error made by approximating the diameter D_M to an arc is very small. Remember that, even for θ as large as 15° (about 0.26 radians) the curved arc and the straight-line approximation to the curved arc differ only by about 1%. In the case of the Moon viewed from the Earth, θ_M is less than 1° so it is a valid approximation to write

$$D_M = L_M \theta_M \qquad (10)$$

Rearranging Equation 10 to find an expression for θ_M, we have

$$\theta_M = \frac{D_M}{L_M} \qquad (11)$$

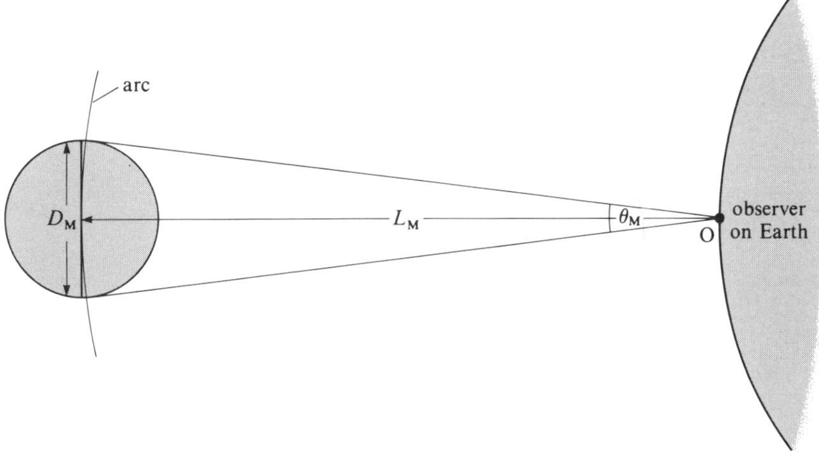

FIGURE 14 If you imagine a circle, centred on O, and passing through the centre of the Moon, the arc corresponding to the angle θ_M will be almost the same length as the diameter of the Moon D_M, provided that θ_M is not too large. Thus, we can write

$$D_M \approx \text{arc} = L_M \theta_M$$

(The subscript M is used to indicate that we are talking about those particular values of D, L and θ that are relevant to the Moon.)

You can see from Equation 11 that the angular size of the Moon is determined by the ratio of the Moon's diameter to the distance of the Moon from the Earth. In Section 3.3 you found the radius of the Moon. Use your value of R_M from Section 3.3 to work out the diameter of the Moon, D_M.*

D_M = ... ± ....km

If you could measure θ_M, you could deduce the distance between the Moon and the Earth. The question is how do you measure θ_M?

□ Look at Figure 15. This should give you a clue as to how to measure θ_M. Can you suggest a way?

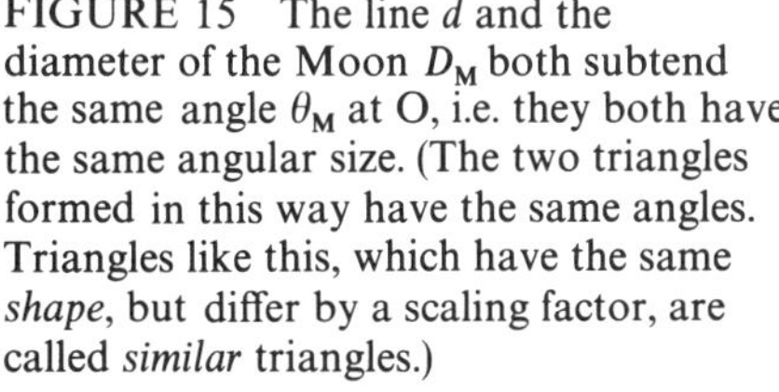

FIGURE 15 The line d and the diameter of the Moon D_M both subtend the same angle θ_M at O, i.e. they both have the same angular size. (The two triangles formed in this way have the same angles. Triangles like this, which have the same *shape*, but differ by a scaling factor, are called *similar* triangles.)

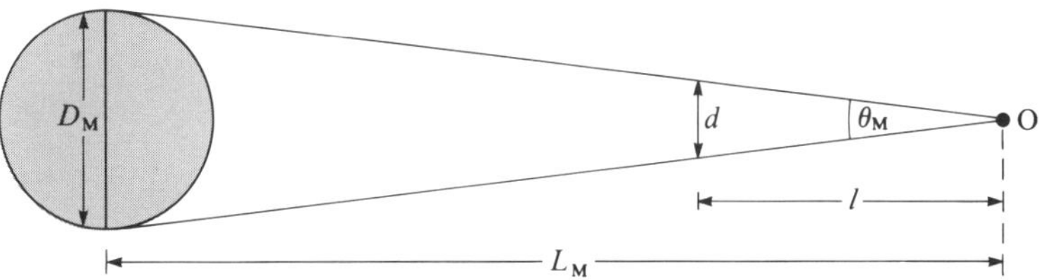

■ θ_M is given by D_M/L_M. But the small-angle approximation can also be applied to the 'arc' d, so that

$$\theta_M = \frac{d}{l} \qquad (12)$$

Thus, if you were to position some object of known diameter d (say) a distance l away from your eye, such that its *apparent* size (i.e. its angular size θ_M) was the same as that of the Moon, you could immediately say that θ_M, the angular size of the Moon, would be given by d/l. Thus, if l can be measured, you can find θ_M (since d is known).

ITQ 14 Your object of diameter d, when placed a distance l away from your eye, must look about the same apparent size as the (full) Moon. Given the hint that the angular size of the Moon is less than 1° (though no information as to how *much* less!), work out what *minimum* diameter an object must have if you want to use it to eclipse the Moon from a distance l of 1 metre.

The only question that remains to be answered is: how could you check that your object of diameter d really is subtending the same angle at your eye as the full Moon? Well, as you'll probably have already guessed by

* The ratio of D_M to the uncertainty in D_M must be the same as the ratio of R_M to the uncertainty in R_M. For example, if the radius of a circle is measured as (2.0 ± 0.1) m, then the diameter of that circle would be quoted as (4.0 ± 0.2) m.

now, the trick is *just* to eclipse the Moon with your object. You then have exactly the arrangement shown in Figure 15. In fact, you don't actually have to work out the value of θ_M in order to find L_M. We have two equations for θ_M:

$$\theta_M = \frac{D_M}{L_M} \tag{11}*$$

and

$$\theta_M = \frac{d}{l} \tag{12}*$$

Because the left-hand sides of these two equations are the same, their right-hand sides must be equal. Therefore:

$$\frac{D_M}{L_M} = \frac{d}{l} \tag{13}$$

Or, rearranging to find an expression for L_M, we have:

$$L_M = \frac{D_M l}{d} \tag{14}$$

Note that all the quantities on the right-hand side of this equation are either known or can be measured.

EXPERIMENT

DETERMINING THE DISTANCE BETWEEN THE MOON AND THE EARTH

TIME

The practical part of this experiment takes about 30 minutes.

NON-KIT ITEMS

piece of dowelling or a broom handle (or some other type of straight rod) at least 1.2 metres long

tape measure or rule at least 1.5 metres long

Blu-Tack or plasticine

KIT ITEMS

Part 1

four plastic discs

FIGURE 16 One possible arrangement for eclipsing the Moon.

SETTING UP THE EXPERIMENT

You now have all the information you need to enable you to find L_M, the distance between the Earth and the Moon. What you need to do is to devise for yourself a simple experimental arrangement that enables you to eclipse the Moon. Figure 16 should set you thinking along the right lines, but you can probably improve on the arrangement illustrated there. After all, it's *your* experiment, so if you think you may be able to get a more accurate estimate of L_M by modifying either the equipment or the technique, then feel free to improvise your own method.

Ideally you require a clear sky and a full Moon to do the experiment properly, but even without these ideal conditions you should still be able to get some kind of result. For instance, if you are careful about matching the curvature of the disc to the curvature of the Moon, you should (with a bit of patience) be able to make the measurements on much less than a full Moon. And although there is not much you can do about cloudy nights, you can, at least, practise your eclipsing techniques on household objects (a standard light bulb at 10 metres is suitable for a test run). You will then be well prepared to take full advantage of the first available cloud-free night.

Tips:

1 For various reasons that are not appropriate to go into at this stage, you will get a more accurate result if you do the experiment in daylight or twilight, rather than on a *dark* night. Alternatively, you could try to work from the window of a well-lit room, rather than doing the experiment outside in the dark.

2 You will find it virtually impossible to block out all the light from the Moon—there will always be some haze around your eclipsing disc. Try instead to match the *curvature* of the disc to the *curvature* of the Moon.

3 You won't find it very easy to decide exactly where the optimum eclipse position is. You faced a similar problem to this in Section 3.3, when you were trying to estimate the radius of the Earth's shadow in the lunar eclipse photograph. There, you found the likely limits, both upper and lower, for this radius. You should do the same sort of thing here; try to estimate the furthest and closest possible eclipse positions, and take the average of these. Remember you should *always* try to assess the uncertainties in every measurement you make.

EXPERIMENT CONTINUED

RECORDING THE RESULTS

The quantities you measure are those on the right-hand side of Equation 14, i.e. the distance l and the disc diameter d. You have several discs, so you need to consider which ones will give you the most accurate answers. You'll probably want at least to *try* them all, and to repeat some of the measurements several times. Don't forget to assess the uncertainties in your values of d as well as in the values of l.

Exactly *how* you record your results is up to you, but you must be able to extract the information at a later stage. The important thing is to write down in your Notebook every aspect of what you did so that you have a permanent record. Don't rely on memory!

WRITING UP A REPORT OF THE EXPERIMENT

You will be asked to 'write-up' this experiment for your first TMA. However, you should *not* attempt to produce the report at this stage. Unit 3 gives you a chance for further practice at such skills as estimating the overall uncertainty when two uncertain quantities (e.g. l and d) are combined. Unit 4 also gives detailed advice about how to construct reports of practical work, with specific reference to this experiment. *You should therefore defer writing up Question 1 of TMA 01 until you have read Unit 4.* Just make sure for now that you have recorded in your Notebook the upper and lower limits to your disc diameter(s) and to the corresponding eye-to-disc distance(s).

But don't be deterred from making a quick calculation, just to see that your data do give a sensible value for L_M: simply substitute your average results for l and d into Equation 14. One final point in connection with that equation: remember that you have to revise your value of R_M (the radius of the Moon) after watching the TV programme. You should, of course, eventually use this revised value in Equation 14, to calculate L_M—don't use the value of D_M you calculated on p. 23.

3.5 THE DISTANCE TO THE SUN

One of the first accurate estimates of the distance of the Moon from the Earth was made by another early Greek 'astronomer', Aristarchus (about 240 BC). And unbelievable though it seems, using a technique very similar to the one you have just used, he actually obtained a result that was within a few per cent of our present-day value! When Aristarchus tried his hand at estimating the distance from the Earth to the *Sun*, however, he got a value that, when compared with the present-day one, was a factor of 20 times too small! Why was he so wrong?

The simple answer to this is that (even today) the distance to the Sun is much harder to estimate than the distance to the Moon, because the Sun is much further away. Another reason for Aristarchus's error being so large was that he used an *indirect* technique. This, though ingenious, had the defect that a small error in the quantity he actually measured led to an enormous error in the quantity he was ultimately trying to find. Follow Aristarchus's reasoning through, and see if you can spot where this huge error creeps in.

Aristarchus argued as follows. The Moon goes through various phases—new Moon, half Moon, full Moon, etc.—as the position of the Moon changes relative to the positions of the Earth and the Sun. (Recall Section 3.3 of Unit 1.) Thus when the Moon appears exactly as a half Moon (i.e. first or last quarter), the sunlight must be striking the Moon at right angles (i.e. 90° or $\pi/2$ radians) to the line of sight of the observer watching the Moon. This situation is illustrated in Figure 17. At this particular time, the angle between the direction of the Moon and the direction of the Sun (labelled ϕ (Greek letter phi) in Figure 17) is measured. This angle is

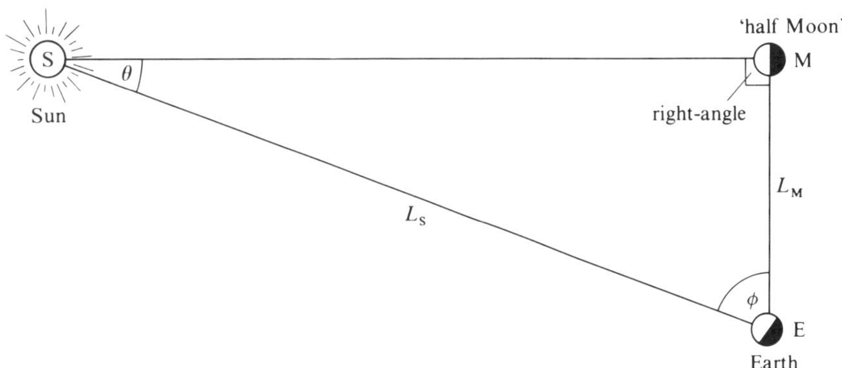

FIGURE 17 Aristarchus estimated L_S, the distance from the Earth to the Sun, by measuring the angle between the Moon and the Sun (the angle labelled ϕ in the diagram) at the moment when the Moon appeared to be exactly a 'half Moon'. He then deduced θ. Since θ was small, he used the equation $L_M = L_S\theta$ (which is just arc = $R\theta$) to find L_S. (The angle θ has been exaggerated in this diagram to make it clearer.) Aristarchus estimated that $\theta \approx 3°$.

almost—but not quite—a right angle (though it is much less than this in the Figure, where the scales are very distorted!).

The Greek mathematicians knew that *the sum of all angles in a triangle is equal to 180°*. So, since the angle at M is a right angle, it follows that:

$$\theta + \phi + 90° = 180°$$

so $$\theta = 90° - \phi \qquad (15)$$

Aristarchus measured ϕ to be about 87°, so he deduced that $\theta \approx 3°$. Now 3° is a small angle, so we can make use of the small-angle approximation for the equation: arc = $R\theta$. Thus we can write:

$$\text{arc} \approx L_M$$

$$R \approx L_S$$

Therefore

$$L_M = L_S\theta$$

or $$L_S = \frac{L_M}{\theta} \qquad (16)$$

ITQ 15 Calculate (using Aristarchus's value of $\theta = 3°$) how many 'Moon-orbit distances' the Earth is away from the Sun. That is, find the ratio ratio L_S/L_M.

The present-day values for the Earth–Sun and Earth–Moon distances (listed on the back cover) give a ratio L_S/L_M of roughly 400. Aristarchus's result is 20 times too small. Why?

The currently accepted value for the angle ϕ is approximately 89.85°. Aristarchus measured this angle as 87°. If we compare Aristarchus's value of ϕ with the present-day one, it is clear (in retrospect) that his measurement was in error by about $(89.85 - 87)° = 2.85°$. An error in measurement of 2.85° in 87° can be expressed in the form of a percentage as:

$$\frac{2.85}{87} \times 100\% \approx 3\%$$

At first sight, this seems quite a small discrepancy, but let us analyse things in a little more detail.

The problem with Aristarchus's method was that he did not use the value of ϕ *directly* to find L_S. As Equations 15 and 16 show, the angle used in the calculations is not ϕ, but θ, which is $(90 - \phi)°$. His value of ϕ was thus $(90 - 87)° = 3°$, when it should have been $(90 - 89.85)° = 0.15°$. Even though Aristarchus's determination of ϕ was only 3% out, his value of θ was a factor of 20 (i.e. 3°/0.15°) too large, and it was this inaccuracy that caused his result for L_S to be a factor of 20 too small.

There's a moral here. Whenever a calculation involves taking the difference between two quantities that are nearly equal, a small percentage error or uncertainty in these measured quantities can produce a very large percentage error or uncertainty in the final result. You will meet similar situations in other experiments later in the Course; in practical work, it is always worth watching out for this kind of problem.

Using the present-day value of $\theta = 0.15^\circ = (2\pi \times 0.15/360)$ radians in Equation 16, we get:

$$L_S = \frac{L_M}{(2\pi \times 0.15/360)} = \frac{L_M}{(\pi/1\,200)} \approx 400\ L_M \tag{17}$$

that is, the Sun is about 400 times farther away from us than is the Moon. Now use the value you obtained for the distance to the Moon (after applying the correction described in the TV programme) and substitute it into Equation 17 to calculate the distance from the Earth to the Sun:

$L_S =$ .... metres

3.6 THE RADIUS OF THE SUN

How big do the Sun and the Moon look to you? Or, to ask the same question more scientifically: How do the apparent sizes of the Sun and Moon compare? Think about this question for a minute. The interesting observation is that the Sun and Moon *look roughly the same size*. However, it's difficult to be precise about this, because the Sun is so much brighter than the Moon. Is there any independent evidence to support this impression that the Sun and Moon have the same angular size?

Actually there is. You've probably seen, at some time or another, photographs of solar eclipses. You've perhaps actually observed annular eclipses (Figure 18a), or even total eclipses (Figure 18b). These provide direct evidence that the Sun and Moon have approximately the same angular size. The Earth–Sun–Moon geometry is shown in Figure 19.

Note: **Under no circumstances should you try to eclipse the Sun with your plastic discs. You will damage your eyes.**

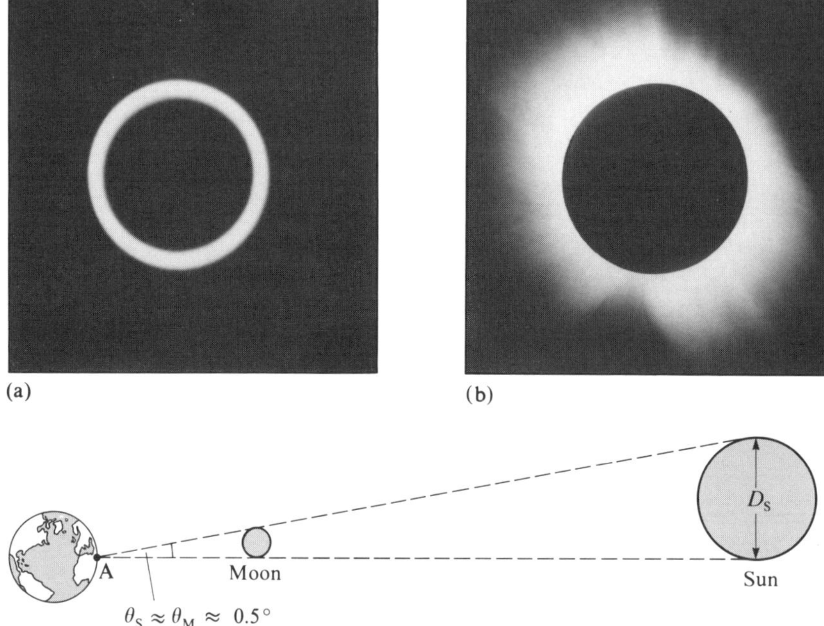

FIGURE 18 (a) A photograph of an *annular* solar eclipse. In this type of eclipse the Moon does not quite block out all of the Sun.

(b) A photograph of a *total* solar eclipse. In this case the basic disc of the Sun is completely obscured by the Moon so allowing observation of the flares in the Sun's 'outer atmosphere'.

FIGURE 19 The Sun and the Moon have approximately the same angular size—they both subtend an angle of about 0.5° at the Earth. This diagram, which is not to scale, shows how a total eclipse is produced at the point A on the Earth's surface. An annular eclipse (like that shown in Figure 18a) is seen when the Moon is slightly further away from the Earth. Two types of eclipse occur because the Moon's orbital radius varies slightly.

Figure 19 shows basically the same arrangement as the one you used in the experiment (Section 3.4) to find the distance to the Moon. The only difference is that here the Moon is eclipsing the Sun, whereas in your experiment a plastic disc was eclipsing the Moon. Nevertheless, the analysis must be identical, that is

$$\theta_S = \theta_M$$

where $\theta_S = \frac{D_S}{L_S}$ and $\theta_M = \frac{D_M}{L_M}$

so that

$$\frac{D_S}{L_S} = \frac{D_M}{L_M} \qquad (18)$$

Or, multiplying both sides of Equation 18 by L_S:

$$D_S = D_M \frac{L_S}{L_M} \qquad (19)$$

You can now use your value for D_M from Section 3.3 (corrected in the light of the TV programme) to estimate D_S. Remember that Equation 17 shows that the approximate value of the ratio L_S/L_M is 400.

D_S = ... metres

Therefore

radius of the Sun, $R_S = D_S/2$ = ... metres

3.7 BRINGING THE RESULTS TOGETHER

You are now in a position to be able to summarize the sizes and distances involved in the Earth–Sun–Moon system by completing Table 4, using your own data whenever possible. Standardize on units by quoting all your results in metres. (Take care to give your values the correct powers of ten.) As your starting point in this Table, you should use the presently accepted value for R_E (the radius of the Earth) of 6.38×10^6 metres. The values you write down should, of course, take account of the correction to R_M discussed in the TV programme and in the TV Notes at the end of this Unit.

In the right-hand column of Table 4, you should make some comment on the accuracy of the result you give. If possible, write down the uncertainty in numerical form, but if this is not possible (because the result is a combination of your data with data taken from the text, for instance) you should try to give some idea of the reliability of the result in words.

ITQ 16 How many Earth radii is the Sun away from the Earth?

TABLE 4 Summary of results

Measurement	Comment on accuracy
R_E, radius of Earth = 6.38×10^6 metres	
R_M, radius of Moon = ... metres	
R_S, radius of Sun = ... metres	
L_M, distance of the Moon from the Earth = ... metres	
L_S, distance of the Sun from the Earth = ... metres	

TANGENT TO A CIRCLE

SUMMARY OF SECTION 3

In this Section, using just a pole in the ground, some photos of the Moon, a length of dowelling, a few plastic discs, and a lot of ingenuity, you have been able to estimate the size of the Earth, the size of the Moon, and the distance between the Earth and the Moon. Quite impressive! Equally valuable, however, is the fact that, in taking and analysing these measurements, you have made use of a number of mathematical ideas and scientific skills, the most important of which are listed below.

1 The circumference C of a circle is related to its radius R (and its diameter D) by the equation $C = 2\pi R = \pi D$.

2 There are two units of angular measure in common usage: radians and degrees (°).

$$2\pi \text{ radians} = 360^\circ = 1 \text{ complete circle}$$

Hence $\quad 1 \text{ radian} = (360/2\pi)^\circ$

and $\quad 1^\circ = (2\pi/360) \text{ radians}$

3 In a circle,

$$\text{arc length} = \text{radius} \times \text{subtended angle (in radians)}$$

i.e. $\quad \text{arc} = R\theta$

4 If θ, in the equation arc $= R\theta$, is small (i.e. less than about 0.26 radians or 15°), then the alternative equation $s = R\theta$, where s is the straight-line approximation to the curved arc, is also true to an accuracy of better than about 1%. This is known as the small-angle approximation.

5 A 'best estimate' for the value of a measurement can generally be found by taking the average of the upper and lower limits of that measurement.

6 Measured quantities should be expressed in the form: best estimate plus or minus the uncertainty in the measurement, and the units must always be given,

i.e. $\quad \text{quantity} = (X \pm x) \text{ units}$

where X and x are numerical values.

4 THE PLANETS

4.1 COPERNICUS'S CONTRIBUTION

Although the Greek astronomers of Alexandria were able to make quite reasonable estimates of the dimensions of the Earth–Sun–Moon system, they certainly didn't know the distances to the planets or to the stars. (The Greeks did realize that there was a difference between the planets and the stars: the planets 'wandered' about relative to the constellations.) The best they could do was to presume that the planets were further away than the Moon, and the stars further away than the Sun and planets. It was not until the beginning of the 16th century when Nicolas Copernicus (1473–1543) developed his theory of planetary orbits—with a stationary Sun at the centre of things (Figure 20)—that it became possible to estimate the *relative* distances to the planets. But once the idea had been mooted of a Sun-centred system, with the planets travelling around the Sun in circular orbits, it became possible—admittedly with some fairly complicated reasoning—to begin calculating the relative radii of these orbits.

For instance, Copernicus deduced the ratio of the radius of the Earth's orbit to that of Venus in the following way. (Recall from Unit 1 that Venus's orbit is closer to the Sun than is the Earth's.) Venus's orbit lies in almost the same plane as the Earth's so that, seen from the Earth, Venus seems merely to swing to and fro, relative to the Sun, first to the left then to the right of it, sometimes passing in front and sometimes behind (Figures 21a and 21b). As you can see from the two diagrams in Figure 21, the extreme right-hand position of Venus's apparent oscillation (i.e. position B) is reached when the line of sight from Earth to Venus *just touches* the circu-

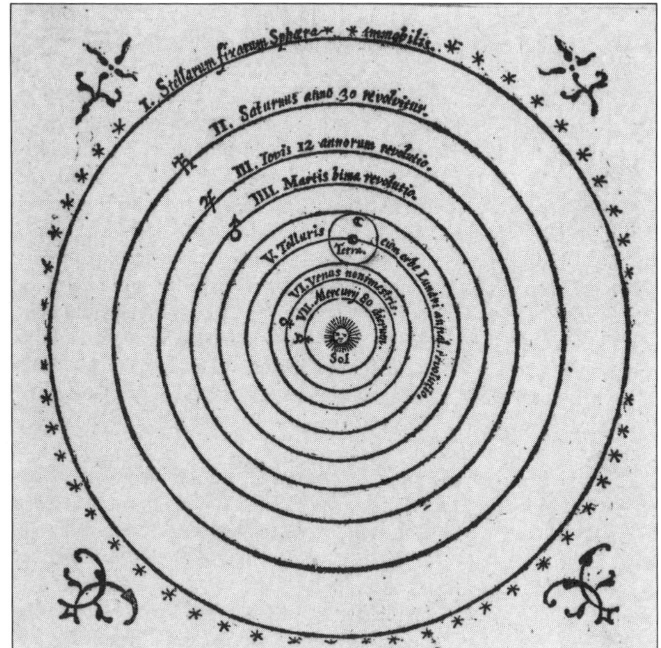

FIGURE 20 The Copernican system of planets with the Sun at its centre. Aristarchus had actually suggested a Sun-centred system back in the third century BC. Unfortunately, he was ahead of his time—tradition and 'logic' were against him. Furthermore, he was not able to make any measurements to support his 'peculiar' theory.

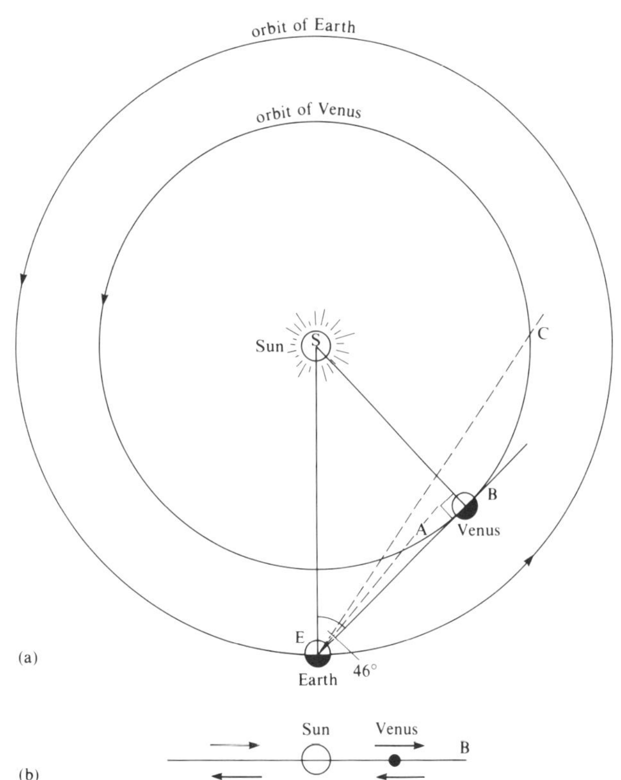

FIGURE 21 (a) The planet Venus orbits the Sun in more or less the same plane as the Earth. Its orbital radius is smaller than the Earth's.

(b) Venus's orbit, as seen from the Earth, appears to oscillate backwards and forwards across the Sun in a straight line.

lar orbit that Venus is assumed to be following. This line EB, which just touches the circle that is Venus's orbit, is said to be a **tangent** to that circle. You can see that a line of sight to a point earlier in the orbit (e.g. EA), or a line of sight to a point later in the orbit (e.g. EC), always corresponds to a smaller angle of deviation from the Sun's direction. The tangent to the circle corresponds to the angle of maximum deviation of the line of sight. Now the Greek mathematicians had shown long before that *a tangent to a circle is always at right angles to the radius of the circle that passes through the tangent's point of contact*. So in Figure 21a we can say that when Venus appears to be at its maximum angular distance from the Sun (i.e. at B in Figure 21b), then the angle at B (denoted as angle EBS) is 90°. If we were also to measure the angle at E (i.e. angle SEB) at this instant of time—which we could do by simple sighting from the Earth—then we could deduce the third angle in the triangle, angle BSE. Copernicus found the angle SEB to be 46°. He therefore deduced, using the fact that the sum of the angles in a triangle is 180°, that the angle BSE (the angle at S) must be 44° (46° + 44° + 90° = 180°).

SINE (SIN)

COSINE (COS)

TANGENT (TAN)

HYPOTENUSE

Once all the angles of the triangle are known, a scaled-down version of this triangle can be drawn. As you may remember from Figure 15, triangles that are identical apart from a scaling factor are called 'similar' triangles. If the Earth–Sun distance is drawn (arbitrarily) 100 mm long, then it turns out that the Venus–Sun distance must be about 72 mm long. (Try it, if you're not convinced.)* Hence the radius of Venus's orbit must be 0.72 times the radius of the Earth's orbit.

Unfortunately, for Copernicus to turn this ratio of orbital radii into an absolute value for the radius of Venus's orbit, he needed an accurate value for the radius of the Earth's orbit (i.e. the value of L_S from Section 3.5). We know this distance quite accurately nowadays, of course, but the best estimate Copernicus had was that determined by the early Greeks—and that value was a factor of 20 out. Consequently, Copernicus got the orbital radius of Venus wrong by this factor.

Needless to say, there was nothing special about Venus. Copernicus repeated similar calculations for the other planets and, although the final *scale* of his Solar System was wrong, he did manage to get the *ratios* between the radii of the planetary orbits roughly correct.

Notice, however, that implicit in Copernicus's calculations was the assumption that the orbits were perfectly circular. We now know that this was not a correct assumption. Because of this, the Copernican measurements did not quite fit the facts. For example, since the orbital periods of the planets were quite well known (from the many years' records of detailed sightings), and since Copernicus claimed to have calculated the relative orbital radii of the planets, it should have been possible to *predict* where a planet would be in the sky (relative to the Earth) at some specified time in the future, or to deduce where it had been some time in the past. It was here that there appeared discrepancies between predictions of the Copernican model and experimental results. Obviously, his model was not satisfactory.

Copernicus spent many years trying to modify his simple model so as to obtain better agreement with the observations. In a way, he succeeded, because he did eventually manage to produce a model in which the agreement between observation and theoretical prediction was very good. But he paid a price for this agreement: his model ceased to be simple! In truth, his modified model of the Solar System, with the Sun at the centre, was just as

* You don't *have* to draw this triangle to find the ratio of two of the sides. Instead you can use the mathematical techniques of *trigonometry*. (See *Into Science*, Module 10, for a more detailed discussion.) In any *right-angled triangle*, the ratio of the lengths of any two of the sides can be related to the angles of that triangle. Three particularly useful ratios are **sine, cosine** and **tangent** (this tangent is *nothing* to do with the tangent to a circle), which are defined thus for the angle θ:

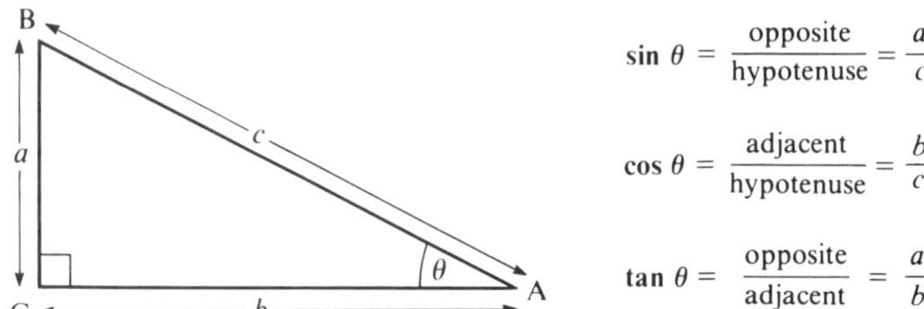

where **hypotenuse** is the name given to the side facing the right angle, *opposite* denotes the side opposite angle θ, and *adjacent* is the name given to the shorter side adjacent to θ.

These quantities have been worked out for all angles between 0° and 90° and tabulated in trigonometrical tables, or programmed into pocket calculators.

So in Figure 21a,

$$\sin 46^\circ = \frac{SB}{SE} = \ldots\ldots\ldots\ldots\ldots\ldots ?$$

(*Answer:* 0.72)

complicated as the 'old-fashioned' geocentric (Earth-centred) models. It was perhaps this fact, more than anything else, that gave the champions of the geocentric system—mainly the philosophical and religious bodies of the time—confidence in the 'rightness' of their case.

4.2 TYCHO BRAHE'S TABLES

The Danish astronomer Tycho Brahe (1546–1601) adopted a quite different approach to that of Copernicus. Rather than invent a model and then try to refine it so as better to fit the available data, he decided to improve the quality and quantity of the data. So he determined to keep a record of observations of the positions of all the planets (five, plus the Earth, were then known) at regular periods throughout the year. In fact, this collection of data became his life's work; he recorded the planetary positions not just for one year, but for more than 20 years. And because of his painstaking development of more and more accurate sighting devices (Figure 22), many of these planetary positions were determined to an accuracy of better than $(1/60)^\circ$ and some to better than $(1/360)^\circ$! So Tycho Brahe left to posterity the most accurate and comprehensive catalogue of 'heavenly activities' that history had so far seen. It is no exaggeration to say that these details formed the basis of future developments in the theory of planetary motion. For one of the most challenging tests that any new theory had to pass was that it had to fit this wealth of data compiled by Brahe.

FIGURE 22 One of the most important sighting instruments in Tycho Brahe's observatory at Uraniborg in Denmark was this huge brass quadrant arc. The arc itself was securely fixed into a western wall, and a south-facing open window was located at the centre of the arc. The empty wall-space inside the arc was decorated with a mural showing Brahe, his dog and his laboratories. In this sketch (taken from Brahe's own book), an observer looking through a pin-hole at F (on the extreme right) is locating the position of a 'star' to an accuracy of better than $(1/60)^\circ$—in fact the scale of the instrument could be read to $(1/360)^\circ$.

ELLIPSE

KEPLER'S FIRST LAW

4.3 KEPLER'S SEARCH FOR REGULARITY

4.3.1 KEPLER THE 'PROBLEM SOLVER'

Johann Kepler (1571–1630), born in Germany, was a generation younger than Tycho Brahe. As astronomers, they were more or less exact opposites. Tycho was a brilliant experimenter and observer. His mechanical ingenuity, as witnessed by his development of numerous observational aids, seemed boundless. Kepler, on the other hand, was a solver of puzzles. He had the kind of mind that delighted in, and was fascinated by, the relationships between numbers, or sizes, or geometrical shapes. So, in some ways, Kepler was the obvious man to tackle the puzzle posed by Tycho Brahe's tables.

Kepler also had an almost mystical belief that there was some mathematical scheme underlying the planetary system. Why only six planets? Why were the planets' orbital radii in the ratio 8:15:20:30:115:195? (These were roughly the relative radii calculated, from Tycho's data, for the Copernican scheme of planets.) Kepler felt sure that there was some sequence to these numbers, and some mathematical explanation for there being only six planets. We know now, of course, that he was wrong. As you will see in Unit 3, the laws of gravitation allow a planet to orbit the Sun at *any* radial distance—so the *ratio* of the orbital radii can have no significance. Furthermore, we now know that there are more than six planets—we have added Uranus, Neptune and Pluto to the list. Nevertheless, after months of work, Kepler did come up with an explanation for the ratio of radii—an explanation based on the geometry of the five regular solids (Figure 23). He was so pleased with his explanation that he published it in a book, copies of which

FIGURE 23 The ratios of the orbital radii of the planets—the right answer ... but the wrong reason. (a) The five regular solids, (b) Kepler's scheme of regular solids (taken from his book).

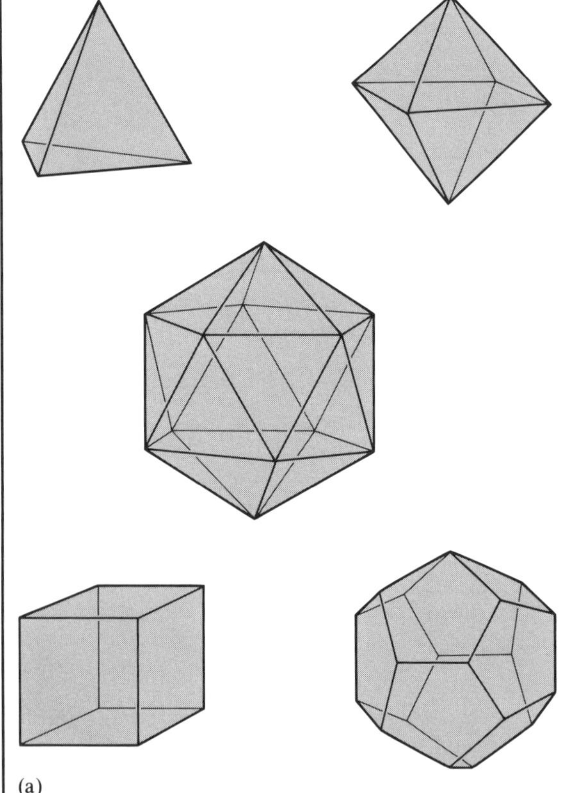

(a)

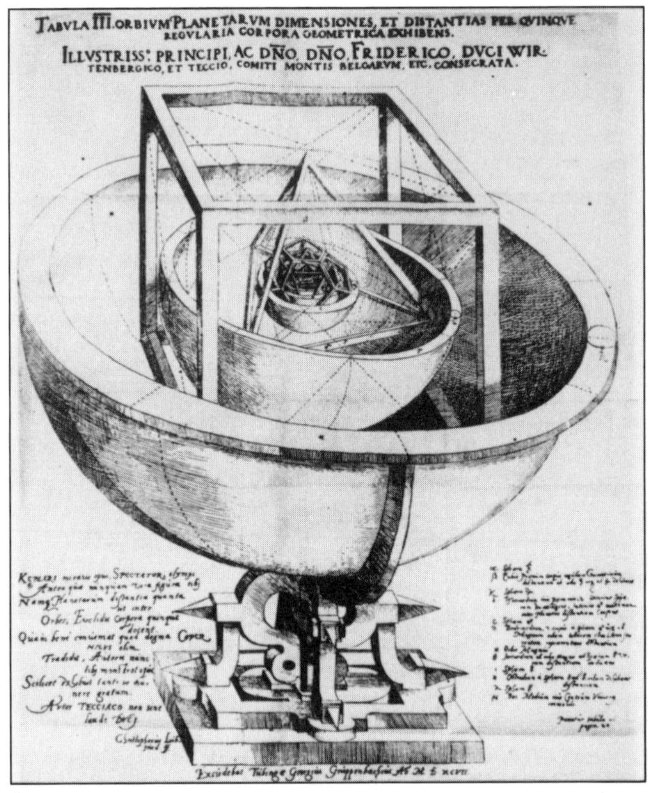

(b)

A *regular solid* is the name given to a geometrical solid that has all its faces identical, regular and plane. Hence a regular solid must have all its edges the same length, all its faces the same shape, all its face angles equal, and all its corners identical. Because of the limitations of three-dimensional space, only *five* such solids are possible (see (a)). Kepler had the idea of 'nesting' one of these solids inside a sphere (so that its corners just touched the sphere), and then nesting a second sphere inside the solid (again so that it just touched), and then nesting a second solid inside the second sphere, and so on, until he had used all five regular solids (see (b)).

By choosing the correct sequence of regular solids he managed to get close agreement between the ratios of the planets' orbital radii, and the ratios of the radii of the 'nested' spheres. Furthermore, since there are only five regular solids, it is only possible to have *six* 'nested' spheres, thus explaining why there were only six planets! **Kepler was wrong**. We now know that this arrangement between sphere radii and planetary orbital radii *was pure coincidence*. We now also know that there are *nine* planets, not six.

he sent to Tycho Brahe and the Italian scientist, Galileo. Both scientists were favourably impressed, and Tycho Brahe invited Kepler to go to Prague to work with him on observations of Mars, 'the difficult planet'. So it was that Kepler became acquainted with the detailed tables drawn up over the years by Tycho. Indeed, the tables were still unpublished when Tycho died, and it fell to Kepler to publish them for him posthumously.

4.3.2 KEPLER'S FIRST LAW

At the time of Tycho Brahe's death, Kepler was deeply involved in a detailed study of the orbit of Mars. Using the data accumulated by Brahe, he tried to fit Mars first into a circular, Sun-centred orbit and then into a circular orbit with the Sun off-centre. Neither worked. Eventually it became clear to him that he would have to plot out, point by point, an accurate and detailed picture of the real orbit of Mars. The problem was not a simple one. All the information was there in Tycho Brahe's tables, but in a scrambled form. The difficulty was that the data gave the apparent position of Mars as seen from a moving Earth.

However, Kepler persevered, and after much calculation he did determine the *shapes* of the orbits of the Earth and of Mars. The Earth's orbit was very nearly circular; indeed, the apparent deviation from circularity could perhaps have been attributable to experimental uncertainty. But Mars was quite different. The orbital path that he had plotted out for this planet was far from being circular; it was quite clearly 'oval' in shape. Kepler failed for several years to guess the exact form of this particular oval shape, but eventually he realized that it was an **ellipse** with the Sun at one focus (Figure 24). He then tested out the ellipse idea (again with the Sun at one focus) on the other known planets in the Solar System. It worked. We now know this result as **Kepler's first law.**

KEPLER'S FIRST LAW

The planets of the Solar System orbit the Sun along elliptical paths. The Sun is at one focus of the elliptical orbits.

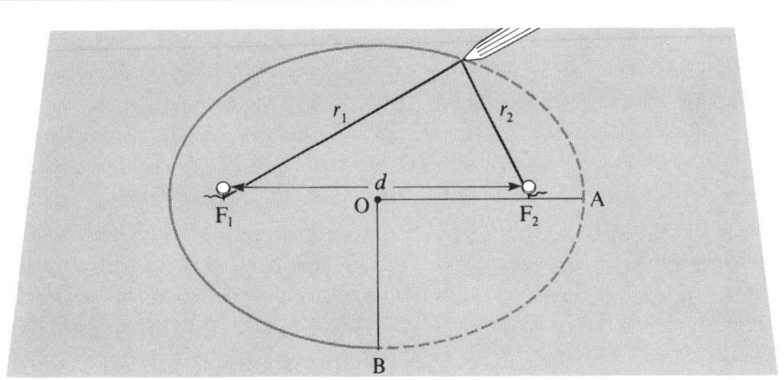

FIGURE 24 Defining an ellipse.

An ellipse is a very easy shape to draw if you have two drawing pins, a piece of string and a pencil. First, fasten the ends of the string to the 'pin-parts' of the two drawing pins. Now press the drawing pins into your drawing surface a distance d apart, where d is less than the length of the string. Take a sharp pencil, and with the tip, extend the string until it is taut. Now draw the curve which the pencil follows when it is moved in such a way as to keep the string taut. This curve is an ellipse; the pins are at the foci of the ellipse. An ellipse can be defined as that curve for which the sum of the distances from the two foci to any point on the curve, is constant, i.e. in the diagram, $r_1 + r_2 =$ constant. The shape of the ellipse can be altered in one of two ways. The distance between the two foci can be changed without changing the length of the string, or the length of the string can be changed without changing the positions of the foci. The distance OA (in the diagram) is known as the semi-major axis, and the distance OB as the semi-minor axis. If the two foci are made coincident (i.e. $d = 0$) the ellipse reduces to a circle.

Kepler's first law says that the planets orbit the Sun along elliptical paths, with the Sun at one focus of the elliptical orbits. The other focus has no significance in the case of planetary motion.

KEPLER'S SECOND LAW

GRAPH

AXES OF GRAPH

4.3.3 KEPLER'S SECOND LAW

Kepler's plot of the orbit of Mars, built up as it was from points separated by equal intervals of time, showed him that the planet moved with an uneven speed around its orbit (Figure 25). He was determined to find some pattern behind this 'unevenness'. He had once, much earlier in his life, published a suggestion that a planet was pushed round the Sun by spokes radiating outwards from the Sun—the force behind the spoke being smaller the longer the spoke. The idea sounds ridiculous today. Planets don't need anything to push them to keep them going. But then, 16th-century ideas of motion were rather confused.

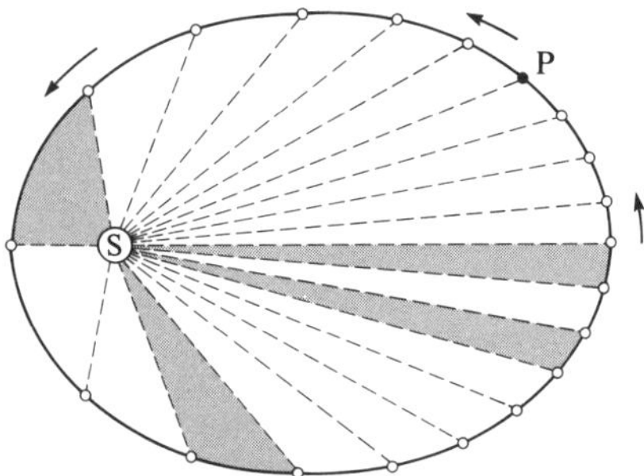

FIGURE 25 The planets orbit the Sun in elliptical paths (with the Sun at one focus of the ellipse). The positions of planet P shown here are separated by equal intervals of time, namely 1/20 of the planet's 'year'.

Nevertheless, this idea led Kepler to spot a very important relationship. He noticed that the speed of the planet's motion was uneven in such a way as to make the 'spokes' from the Sun sweep out exactly equal areas of space in equal times. All the shaded areas in Figure 25 are equal. So, if the spoke is a short one (as the planet passes near the Sun), then the speed of the planet must be large to compensate for this. Conversely, the speed is much less when the spoke is long. We now know that this holds true for all the planets: it is a general law, now called **Kepler's second law.**

KEPLER'S SECOND LAW

The 'spoke' joining the Sun to a planet sweeps out equal areas in equal times.

Notice that Kepler did not explain *why* this should be the case. He merely observed that it *was* the case. He had discovered the pattern—but he had not explained it. So, although his second law was a very useful tool for predicting the future positions of planets, it did nothing to explain the 'mysteries of space'. In fact, Kepler's second law can be explained using Newton's (more general) law of gravitation, which you will be learning about in Unit 3.

4.3.4 THE RELATIONSHIP BETWEEN ORBITAL PERIOD AND ORBITAL RADIUS

There was another 'planetary numbers' puzzle that had been worrying Kepler: what was the relationship, if any, between a planet's 'year' (i.e. the time it takes to complete one orbit round the Sun—its orbital period) and its orbital radius? He had all the data from Tycho Brahe's records, but he had not yet spotted the pattern in the data. There was one obvious regularity—the planets with the largest orbital radii* had the longest orbital times. But this is what you would expect—these planets have further to travel in one of their 'years'. Kepler was sure there must be a more definitive relationship hidden away in the data.

TABLE 5 Planetary data

Planet	Orbital period T (in units of Earth-years)	Orbital radius R (in units of Earth-orbital radii)
Mercury	0.24	0.39
Venus	0.62	0.72
Earth	1.00	1.00
Mars	1.88	1.52
Jupiter	11.86	5.20
Saturn	29.46	9.54

Table 5 shows the sort of planetary data Kepler had to work on. (Actually, these are modern data, which are slightly more accurate than those used by Kepler.) Can *you* see any relationship between T and R?

4.3.5 PLOTTING A GRAPH

If you didn't manage to spot the relationship between T and R by playing around with the numbers in Table 5, you're in good company. It took Kepler a long time to see the solution—the problem is not trivial! Nowa-

* Since the orbits are ellipses, the word 'radius' here should be interpreted to mean 'average radius'. Strictly speaking, R, as used in Kepler's laws, is the longer (semi-major) axis of the ellipse (see Figure 24).

days, however, we have a much better way of approaching it than Kepler did: we can plot a **graph**.*

We now realize that if an orbital radius of 0.39 Earth-orbital radii corresponds to an orbital time of 0.24 Earth years, and an orbital radius of 0.72 Earth-orbital radii corresponds to an orbital time of 0.62 Earth years, and so on, the easiest way to show this correspondence is on a two-dimensional diagram, in which one quantity is plotted in one direction and the second quantity is plotted in a direction at right angles to the first. Then, if there is a mathematical relationship between the two quantities, we should expect all the points to lie on a smooth curve; whereas if there is no fixed relationship, we should expect the points to be scattered with no discernible pattern.

ITQ 17 Can you think of a good argument for why a fixed relationship between two quantities should give rise to a smooth curve?

Try drawing this graph now on the grid provided in Figure 26**. The **axes** of the graph have been labelled to help you. (For the moment, just take note of the way the symbols and the units of measurement are used in the labelling; we will come back to the reason for doing it like this shortly.) You will perhaps find it easiest to plot the point corresponding to the largest values of R and T first (i.e. those for Saturn). Look along the horizontal axis until you find the value 9.54 Earth-orbital radii. Draw a faint vertical line at this value. All points on this line must have a value $R = 9.54$ Earth-orbital radii. You also know that this value of R corresponds to only one particular value of T, namely $T = 29.46$ Earth years. So look up the vertical axis until you come to this value of T. Draw a faint horizontal line at this value of T. All points on this line must have the same value of T. So, it must follow that where your faint vertical and horizontal lines cross, R equals 9.54 Earth-orbital radii, and at the same time T equals 29.46 Earth years. That is, this single point represents both the pieces of information corresponding to the planet Saturn.

Now do the same thing for all the other planets in Table 5. After you've plotted a few points, you'll probably find that you don't have to draw the intersecting lines any more—you'll be able to locate their point of intersection by eye. Incidentally, because of the scale of the axes in Figure 26, you'll find that the points corresponding to Mercury and Venus are cramped up near the origin (i.e. the point $R = 0$, $T = 0$). Yet if we expand the scale, the point corresponding to Saturn would go off the page. So, we have chosen the best compromise.

Do your points appear to lie on a smooth curve?

If not, then check to see if any of the points are plotted incorrectly. A wrongly plotted point is often clearly displaced from the curve that can be drawn through the other points. If your points lie roughly in a curve, then draw it in. Do *not* join the points up by straight lines—try to estimate the *smoothest* fit to the points, even if this means just missing one or two of them. Your curve should have no sharp kinks in it.

ITQ 18 If you, as a modern astronomer, were unexpectedly to find a planet orbiting the Sun in an orbit with an average orbital radius of 3.0 Earth-orbital radii, would you feel confident about predicting the orbital period of the planet using only the curve you have sketched in Figure 26? What do you think this orbital period would be?

* It is hard for us to realize nowadays (when graphs are such commonplace devices for displaying data) that the whole concept of graphical representation *post*dated Kepler. In fact, we owe this invention to the mathematician and philosopher, René Descartes (1596–1650).

** If you have any difficulty in plotting the graph, refer to *Into Science*, Module 7, for further assistance.

EXTRAPOLATION

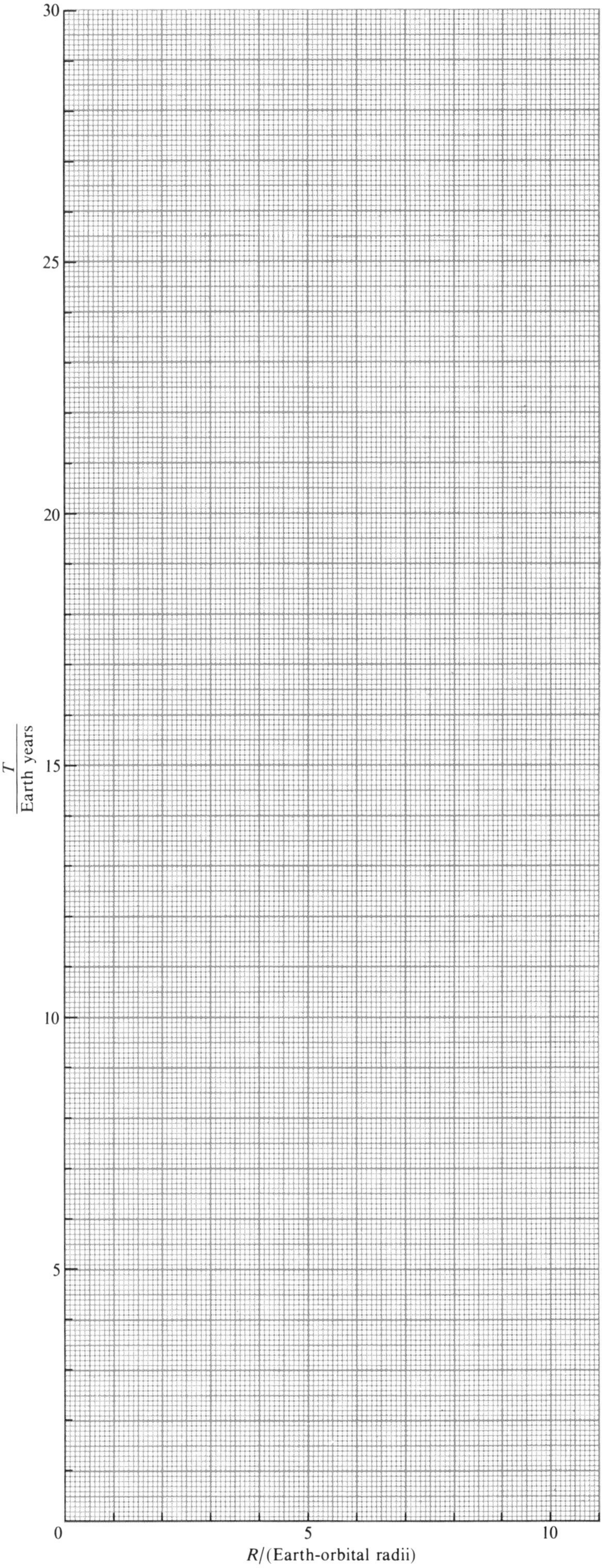

TABLE 5 Planetary data

Planet	Orbital period T (in units of Earth-years)	Orbital radius R (in units of Earth-orbital radii)
Mercury	0.24	0.39
Venus	0.62	0.72
Earth	1.00	1.00
Mars	1.88	1.52
Jupiter	11.86	5.20
Saturn	29.46	9.54

FIGURE 26 Plot points showing the correspondence between R and T for each of the planets given in Table 5 (repeated above). Note that the units of T and R are given below the division line.

ITQ 19 Did you have any trouble in getting your smooth curve to pass through the point corresponding to the Earth? What does this tell you about the Earth?

4.3.6 KEPLER'S THIRD LAW

The graph you have plotted in Figure 26 does show the relationship between T and R, but in a somewhat limited kind of way. Suppose, for instance, that you were to discover a planet with an orbital radius of 39.44 Earth-orbital radii (the value of R for the most recently discovered planet, Pluto). Could you deduce the value of T for this planet from your Figure 26 graph? The problem is that you would have to extend the graph (a process known as **extrapolation**) quite considerably beyond the range covered in Figure 26. This is tricky. Obviously, you would do your best to project the shape of the curve to higher values of T and R but, as you can probably imagine, without knowing the 'theoretical form' of the curve (i.e. without knowing the mathematical equation that relates T to R), your projection could be wildly inaccurate. This is precisely the sort of problem that faces social scientists, or governments for that matter, when they try to predict future trends in birth-rate, or unemployment, or inflation, without any knowledge of the mathematical relationships that describe these things. However, with regard to the planets, there *is* a mathematical relationship between T and R, which Kepler eventually found.

It is asking a bit much to expect you to repeat Kepler's discovery within your week's work on this Unit! However, given the relationship, you can at least verify that it really does work. What Kepler said was that if you take the period of a planet's orbit and square it and then divide the result by the cube of the radius of the planet's orbit, you will always get the same value; that is:

$$T^2/R^3 = \text{constant} \qquad (20)$$

In fact, if you work in units of Earth years and Earth-orbital radii (as in Table 5), the constant you get has a numerical value of 1.00.

Table 6 repeats the data shown in Table 5, but it also has columns for T^2 and R^3 and one labelled T^2/R^3. These three additional columns have been filled in for Mercury and the Earth; the rest have been left for you to complete yourself. Before you do this, however, there are a couple of points to notice.

The first of these concerns the way the columns have been headed. The symbol for each quantity (or combination of quantities) is *divided* by the units in which that quantity (or combination) is measured. As a result, it is possible to enter the data in the body of the Table just as pure numbers: in the case of Venus, for example, the orbital period T is 0.62 Earth years and this is equivalent to saying that:

$$T/\text{Earth years} = 0.62$$

You have already met this kind of labelling system once—on the axes of the graph in Figure 26.

The second important point concerns the number of digits quoted for each quantity. Look at the data for Mercury. R has been measured in Earth-orbital radii to two digits:

$$R = 0.39 \text{ Earth-orbital radii}$$

If you use a calculator to calculate the value of R^3, you will get

$$0.39 \times 0.39 \times 0.39 = 0.059319$$

However, since the original value of R was only accurate to two digits, there is no justification for giving a value of R^3 with greater accuracy; in this context, some of the digits on the calculator display are quite simply meaningless. The value of R^3 entered in Table 6 is therefore

$$R^3 = 0.059 \text{ (Earth-orbital radii)}^3$$

KEPLER'S THIRD LAW

PROPORTIONALITY

CONSTANT OF PROPORTIONALITY

TABLE 6 Testing Kepler's third law

Planet	T Earth years	R Earth-orbital radii	T^2 (Earth years)^2	R^3 (Earth-orbital radii)^3	T^2/R^3 (Earth years)^2/(Earth-orbital radii)^3
Mercury	0.24	0.39	0.058	0.059	0.98
Venus	0.62	0.72			
Earth	1.00	1.00	1.00	1.00	1.00
Mars	1.88	1.52			
Jupiter	11.86	5.20			
Saturn	29.46	9.54			

ITQ 20 Using your calculator, fill in the remaining gaps in Table 6. Are your results for the right-hand column consistent with Equation 20?

The relationship expressed in Equation 20 is now known as **Kepler's third law**.

KEPLER'S THIRD LAW

The square of a planet's orbital period divided by the cube of that planet's orbital radius is a constant or

$$T^2/R^3 = \text{constant} \qquad (20)^*$$

In ITQ 20 you found that the numerical value of the constant is 1.00. This, however, was only because the planetary data were expressed in units of Earth years and Earth-orbital radii. Had we used any other units, the constant would not have been unity, but Kepler's third law (Equation 20) would nevertheless have held true.

4.4 URANUS, NEPTUNE AND PLUTO

Since Kepler's day we have discovered three more planets in the Solar System—Uranus, Neptune and Pluto, and it is interesting to see whether Kepler's third law holds true for these planets as well. After all, Kepler's relationship might not be the only one that satisfies the data relating to the inner six planets: there might be an alternative formula that would also fit the series of numbers. So a seventh, eighth and ninth planet provide a way of testing the third law.

TABLE 7

Planet	R/(Earth-orbital radii)
Uranus	19.14
Neptune	30.20
Pluto	39.44

ITQ 21 Table 7 shows the orbital radii of the three planets Uranus, Neptune and Pluto. What orbital periods does Kepler's third law predict for these planets?

4.5 PROPORTIONALITY

Kepler showed that the relationship between T and R is:

$$T^2/R^3 = \text{constant} \qquad (20)^*$$

or, multiplying both sides of this equation by R^3:

$$T^2 = \text{constant} \times R^3 \qquad (21)$$

What Equation 21 says is that any value of T^2 can be found by multiplying the corresponding value of R^3 by a fixed constant; that is, all values of T^2 are in **proportion** to the corresponding values of R^3. Thus, if R^3 is doubled, T^2 is also doubled; if R^3 is multiplied by four, the corresponding value of T^2 would also be four times bigger.

In all the examples encountered so far in this Unit, the constant in Kepler's third law has always been 1.00, in which case Equation 21 can be reduced to:

$$T^2 = R^3 \tag{22}$$

Remember, though, that this particular situation exists simply because we have chosen to express T and R in units of Earth years and Earth-orbital radii respectively (both of which are 1.00 for the Earth, of course). With different units, the constant might have been different, say 0.2 or 5 or 7.3. This would not have made any real difference, however: the T^2 values would still have been in proportion to the R^3 values. Only the so-called **constant of proportionality** would have changed.

Proportionality relationships occur so frequently in science that we use a special symbol $\propto$ as shorthand for 'is proportional to'. For Kepler's law we would say T^2 is proportional to R^3, and write this as

$$T^2 \propto R^3$$

This must be exactly equivalent to the expression

$$T^2 = kR^3 \quad \text{(where } k \text{ stands for the constant of proportionality)}$$

This is an important rule: *to convert a proportionality into an equality, a constant—called the constant of proportionality—must be included in the equation.*

ITQ 22 In Section 2 you saw that the dimensions on both sides of an equation must be the same. Kepler's third law says that:

$$T^2 = \text{constant} \times R^3$$

The dimensions of T^2 on the left-hand side of this equation are 'time squared'. The dimensions of R^3 are 'length cubed'. So how can this equation be correct?

SUMMARY OF SECTION 4

In this Section you have once again been introduced to two kinds of material: factual information relating to the motion of the planets in their orbits around the Sun, and the skills and techniques needed to analyse these planetary orbits. The information relating to the planets is encapsulated in Kepler's three laws:

1 Kepler's first law states that each planet orbits the Sun along an elliptical path, with the Sun at one of the foci of the ellipse.

2 Kepler's second law states that the imaginary line joining the Sun to an orbiting planet sweeps out equal areas in equal times.

3 Kepler's third law states that, for any orbiting planet, the ratio of T^2 to R^3 (i.e. orbital period squared to orbital radius cubed) is a constant.

In working through this Section you have also gained experience in some of the mathematical techniques that are particularly useful in analysing scientific data, notably those outlined below.

(i) It is generally possible to illustrate the relationship between two quantities by drawing a graph. One of the quantities is plotted along the vertical axis, and the other quantity along the horizontal axis. In general, a graph representing a *fixed relationship* between two quantities will take on the form of a smooth curve rather than a set of points with no obvious pattern.

(ii) An expression of proportionality can be converted into an *equality* (i.e. an equation) by introducing a constant of proportionality. For example, if $x \propto y^2$, then $x = ky^2$ where k is a constant. The *dimensions* of a constant of proportionality must be such as to retain dimensional balance within the equation. For example, if $x = ky^2$, then $k = x/y^2$, and hence the dimension of k must be equal to the dimensions of x divided by the dimensions of y^2.

FIGURE 27 Galileo's telescope. It was probably modelled on the telescope design patented by Hans Lippershey in Holland in 1608.

5 THE MOONS OF JUPITER

5.1 THE WORK OF GALILEO

The other important figure contributing to the science of astronomy at the beginning of the 17th century was the Italian, Galileo Galilei (1564–1642). Indeed, it could be said that it was Galileo who laid the foundations of modern observational astronomy by recognizing that the then newly-invented telescope—which made distant objects appear closer, and therefore larger—could be used to advantage in the study of the heavens. His first telescope magnified objects by a factor of only three, but with patience and perseverance he eventually constructed a satisfactory instrument with a magnification of $\times 30$ (Figure 27). With this new instrument he saw, for the first time, the planets not as points of light, but as luminous discs. The 'stars' were still just points in the sky (obviously much further away than the planets) but through the telescope they were brighter, further apart, and above all, far more numerous.

Perhaps Galileo's most important astronomical discovery was made with this telescope on the night of 7 January 1610. He was studying the region of the sky near the planet Jupiter when he noticed three small new 'stars'. These 'stars', together with Jupiter itself, seemed to form a straight line; one of these 'stars' was to the west of Jupiter, the other two to the east. Although he found this straight-line effect sufficiently interesting to make a sketch of the pattern, he did not think there was anything particularly strange about it. He assumed that these 'stars' were simply three more fixed and distant stars that his new telescope had brought within his view.

The surprise came the following night when he was again scanning the sky near Jupiter. The three 'stars' were still there, but now all three were to the west of Jupiter, and positioned more closely together than before. His first thought was that this shift was caused by the motion of Jupiter relative to the Earth—though he did feel that the size of the shift was surprisingly large to have taken place in only 24 hours. But then he realized that, compared with all the other stars, the shift was *in the wrong direction*—Jupiter would have had to be going the wrong way round its orbit! His curiosity was aroused. He decided to watch the 'stars' every night, and keep a record of their positions. Figure 28 shows Galileo's estimate of the positions of these 'stars' over the period 7 January to 15 January 1610. The night of 9 January must have been very frustrating for him—the sky was cloudy, and he could not see the 'stars' at all. But, on the very next night, he found the 'stars' had moved back to the east of Jupiter. This convinced him that the 'stars' themselves must be moving, thus indicating that they were not really stars after all.

What Galileo had actually found were four of the moons of Jupiter.

5.2 JUPITER'S MOONS AND KEPLER'S THIRD LAW

Kepler quickly realized that Jupiter, together with its moons, formed a kind of small-scale model of the Solar System. So, if his third law applied to the Solar System, why not to the Jupiter system also? Table 8 shows the orbital periods and orbital radii of the four innermost moons of Jupiter (in metric units this time).

TABLE 8 Orbital periods and radii for four of Jupiter's moons

moon	$\frac{T}{\text{hours}}$	$\frac{R}{\text{km}}$	$\frac{T^2}{(\text{hours})^2}$	$\frac{R^3}{(\text{km})^3}$	$\frac{T^2/R^3}{(\text{hours})^2/(\text{km})^3}$
Io	42.4	4.22×10^5	1.80×10^3	7.52×10^{16}	
Europa	85.2	6.71×10^5	7.26×10^3	3.02×10^{17}	
Ganymede	171.7	10.71×10^5	2.95×10^4	1.23×10^{18}	
Callisto	400.5	18.84×10^5	1.60×10^5	6.69×10^{18}	

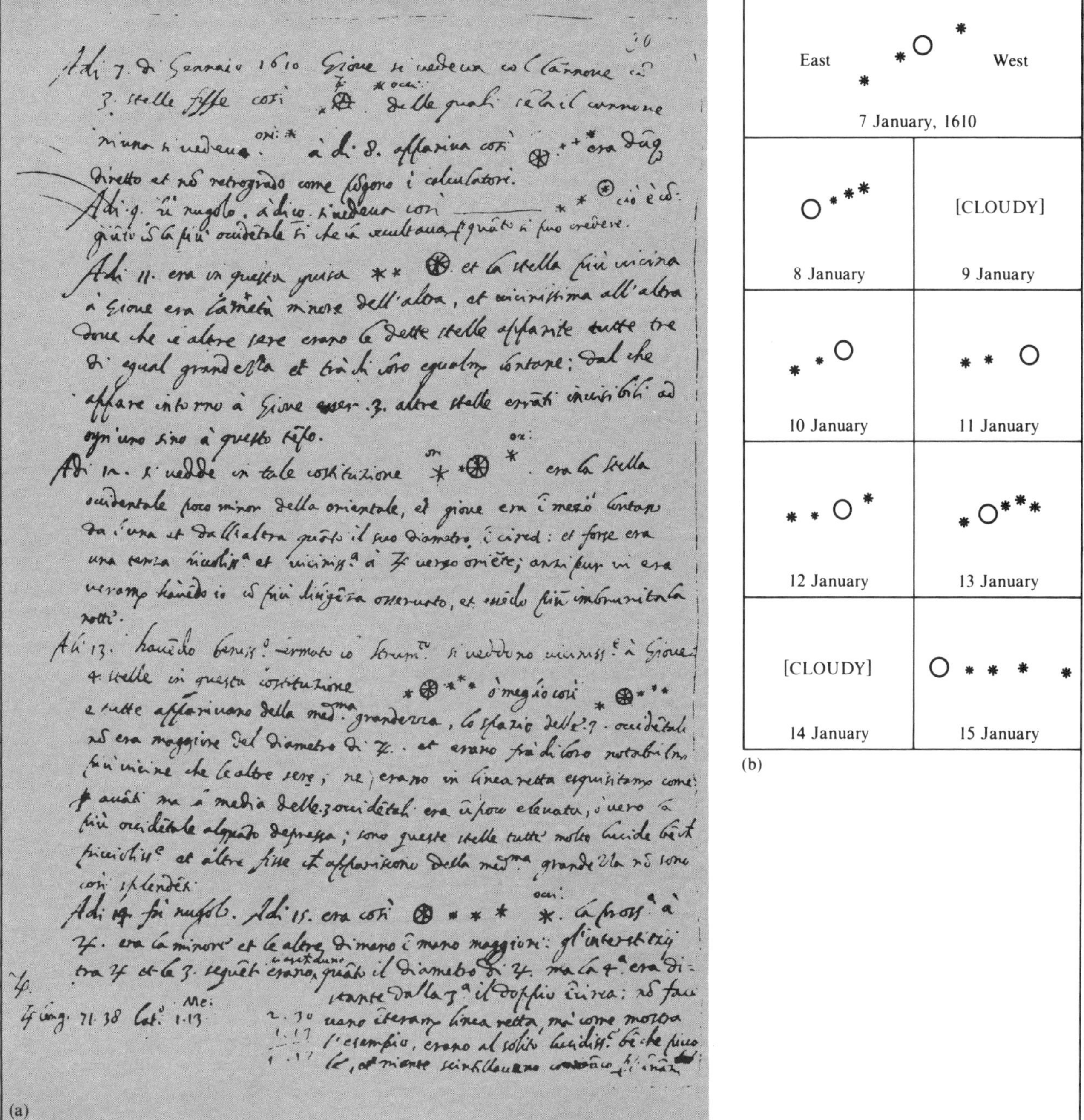

FIGURE 28 (a) A facsimile of the page from Galileo's handwritten notebook showing the positions of Jupiter's moons from 7 January to 15 January 1610.

(b) A summary of the moons' positions each night. The moons' orbits lie almost in the plane containing the line-of-sight from the Earth to Jupiter. Consequently, the moons are frequently in front of, or behind, Jupiter; that is why all four moons are not always in view simultaneously.

ITQ 23 To save you some tedious calculation, the values of T^2 and R^3 have been filled in on Table 8. Using your calculator, complete the right-hand column. Does Kepler's third law apply to Jupiter's moons?

In fact, Kepler's third law applies not just to the Solar System and the moons of Jupiter, but to *any* 'quasi-planetary' system.

We have already seen (Table 6) that, in the case of the planets of the Solar System, the constant of proportionality for Kepler's law is 1.00 in units of (Earth years)^2/(Earth-orbital radii)^3. To be able to compare this constant with the one obtained for Jupiter's moons, we need to convert it to units of (hours)^2/(km)^3.

To make the conversion, we use the facts that

$$\text{one Earth year} = (365\tfrac{1}{4} \times 24)\ \text{hours}$$
$$= 8\,766\ \text{hours}$$

and

$$\text{one Earth-orbital radius} = 1.50 \times 10^8\ \text{km}$$

Therefore

$$1.00\,\frac{(\text{Earth years})^2}{(\text{Earth-orbital radii})^3} = \frac{1.00 \times (8\,766)^2\ \text{hours}^2}{(1.50 \times 10^8)^3\ \text{km}^3}$$
$$= 2.28 \times 10^{-17}\ (\text{hours})^2/(\text{km})^3$$

So, although Equation 20 applies to both the moons of Jupiter and the Solar System, the constant of proportionality is different in the two cases. The difference is approximately a factor of 1 000 (i.e. $10^{-14}/10^{-17}$).

You might guess that the change of constant has something to do with the change of orbital-centre, from the Sun to Jupiter—and you would be right. However, the full explanation is quite a long one and will take us forward by about half a century from the days of Galileo and Kepler to the time of Sir Isaac Newton.

SUMMARY OF SECTION 5

Kepler's third law has been shown to apply not only to the planets in the Solar System, but also to the moons of Jupiter. In fact it is valid for any system of bodies orbiting a common 'centre'. The underlying reason for the universal nature of Kepler's third law was first appreciated by Newton, whose work is the subject of Unit 3.

6 TV NOTES: MEASURING—THE EARTH AND THE MOON

This programme re-examines Eratosthenes' method for determining the radius of the Earth, and the assumptions underlying the use of a lunar eclipse photograph to determine the radius of the Moon. In addition, it may give you a feel for the trials and tribulations—and the fun and enjoyment—of doing practical work in science. As you'll see, experiments don't always go as planned—even for OU academics!

The programme opens at the Bishop Walsh School in Sutton Coldfield where, with the cooperation of some of the pupils, a modern version of Eratosthenes' experiment had been set up. The school is located almost exactly 1.5° of longitude west of Greenwich; so, allowing the extra hour for British Summer Time, it was calculated that local noon would be at 1.06 p.m. However, rather than just taking one measurement of the length of the shadow at exactly local noon, the experimenter decided to follow the movement of the end of the shadow over a period of about 30–40 minutes either side of local noon (Figure 29). If the length of the shadow were to remain constant over this period, then the line of marker-pins would follow the contour of the base board. However, the length of the shadow is *not* constant, because, of course, the Sun ascends higher in the sky during the morning, and is at its culmination point exactly at noon. Hence the shadow length is shortest at noon. The line of marker-pins obtained during the experiment (Figure 29) clearly demonstrates this point.

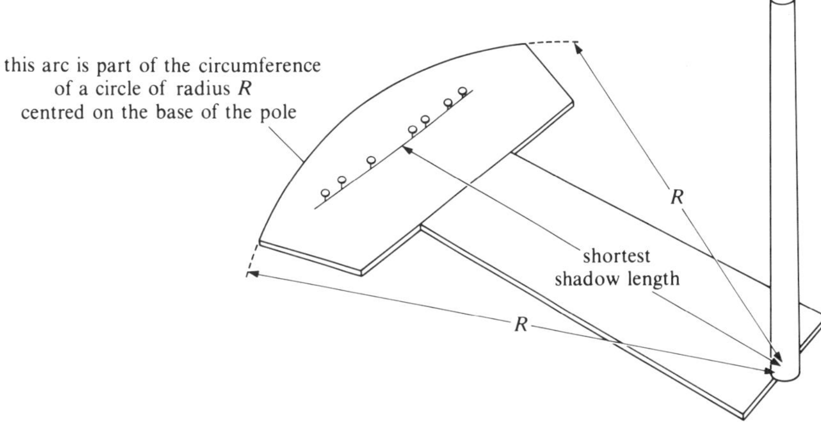

FIGURE 29 The shadow of the pole, cast on the base board, is shortest at noon.

The results of the experiment are summarized below.

At Sutton Coldfield
The length of the shadow at noon = 178 cm
The height of the pole = 318 cm

Hence the angle θ_s between the Sun's rays and the local vertical at Sutton Coldfield = 29.2°.

On the same day, the pupils of Peterhead Academy in Scotland (Peterhead is on the same line of longitude as Sutton Coldfield) found that the angle θ_p between the Sun's rays and the local vertical at Peterhead = 34.4°.

We know that the distance between Bishop Walsh School and Peterhead Academy is 552 km. The difference in angle at the two schools is $\theta_p - \theta_s = 34.4° - 29.2° = 5.2°$. Hence, an angle of 5.2° carries an arc length of 552 km.

In the Eratosthenes experiment, it is assumed that the light coming from the Sun reaches the Earth as parallel rays. However, because the Sun is not a point source of light, this assumption is not quite correct. The consequence of this for the Eratosthenes experiment is that the end of the shadow is somewhat fuzzy; this may have introduced an error of about 1% in the measurement of the length of the shadow. (*Note:* the fuzziness is nothing to do with the cloudiness of the weather.) However, when the shadow is cast over very large distances (as, for example, when the shadow of the Earth is cast on the Moon), this parallel-ray assumption can lead to quite erroneous results. Figure 30 shows how the shadow region behind the Earth is conical in shape.

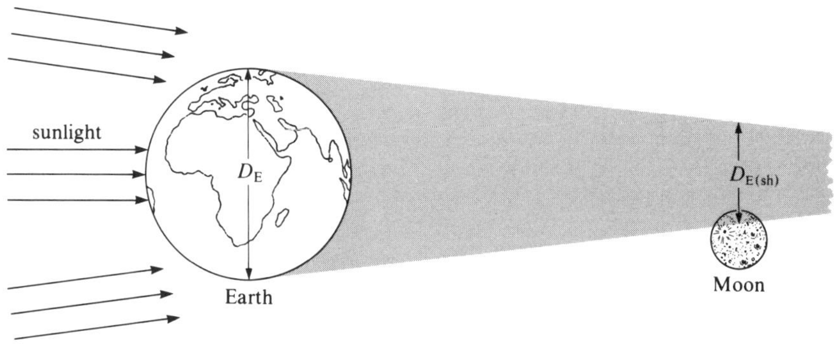

FIGURE 30 The diameter of the Earth's shadow cast on the Moon $D_{E(sh)}$ is smaller than the diameter of the Earth D_E. (Not drawn to scale.)

As explained in the programme, Aristarchus reasoned that the shadow narrows by one Moon diameter over a distance of one Moon orbital radius. That is, the diameter of the Earth's shadow $D_{E(sh)}$* cast on the Moon (i.e. one moon orbital radius away) during a lunar eclipse is less than the diameter of the Earth D_E by an amount equal to the diameter of the Moon D_M; that is:

$$D_{E(sh)} = D_E - D_M$$

Dividing by D_M:

$$\frac{D_{E(sh)}}{D_M} = \frac{D_E}{D_M} - 1$$

or

$$\frac{D_{E(sh)}}{D_M} + 1 = \frac{D_E}{D_M} \qquad (23)$$

In Section 3.3 you measured the ratio of the radius of the Earth's shadow cast on the Moon to the radius of the Moon, $R_{E(sh)}/R_M$, which of course is the same as $D_{E(sh)}/D_M$. But in order to find D_M (knowing D_E) what you really want is the ratio D_E/D_M. You are now in a position to calculate this ratio using Equation 23 and hence to correct your value for D_M (and therefore R_M).

The programme closes with a brief look at the way in which modern laser lunar-ranging techniques allow scientists to make very precise measurements of the distance to the Moon.

OBJECTIVES FOR UNIT 2

Having completed your work on this Unit, you should be able to:

1 Explain the meaning of, and use correctly, all the terms flagged in the text.

2 Estimate upper and lower limits to a measurement and hence calculate a 'best value' by taking the average of the upper and lower limits. (*Experiment and ITQ 11*)

3 Calculate the upper and lower limits to a quantity that is equal to the ratio of two other quantities when the error or uncertainty in one of these quantities is much greater than that in the other. (*ITQ 12*)

4 Appreciate that the percentage error or uncertainty in the difference of two nearly equal quantities can be very large, even though the percentage error or uncertainty in these individual quantities is small.

5 Compile and understand Tables of data. (*ITQs 20 and 23*)

6 Plot a graph and interpret data presented in graphical form. (*ITQs 17–19*)

7 State Kepler's laws of planetary motion; use Kepler's third law to calculate any one of the quantities T, R or k, given the other two. (*ITQs 21 and 23*)

8 Explain how Eratosthenes' method for measuring the circumference (and hence the radius) of the Earth can be adapted so as not to require the Sun to be directly overhead at either of the two locations. (*TV*)

9 Explain why the assumption that the light from the Sun reaches the Earth in the form of parallel rays is only approximately true. (*TV*)

* Denoted by D_S in the TV programme; $D_{E(sh)}$ is used here since D_S is used earlier in this text to denote the diameter of the Sun.

10 Estimate the radius of the Earth from the radius of the Earth's shadow cast on the Moon during a lunar eclipse.

11 Rearrange mathematical equations involving addition, subtraction, multiplication, division, squares, square roots and cubes; calculate the value of a quantity from such an equation.

12 Handle reciprocals, fractions and proportions between quantities. (*ITQs 3, 20 and 23*)

13 Convert an expression of proportionality into an expression of equality by introducing a constant of proportionality.

14 Use the powers-of-ten (scientific) notation. (*ITQs 1 and 2*)

15 Use the SI units of metres, kilograms and seconds, and multiples and fractions of these units; convert physical quantities from one set of units to another, possibly using standard prefixes and/or powers of ten. (*ITQ 2*)

16 Use, appropriately, the order of magnitude symbol. (*ITQ 2*)

17 Explain what is meant by the dimensions of a physical quantity; calculate the dimensions of such a quantity given an equation relating that quantity to others of known dimensions. (*ITQs 8 and 22*)

18 Calculate the radius (or diameter) of a circle given its circumference, or vice versa. (*ITQ 4*)

19 Convert an angle from units of radians to units of degrees, and vice versa. (*ITQs 6 and 7*)

20 Use the equation arc $= R\theta$ and the small-angle approximation when appropriate; hence relate the angular size of an object to its real size and its distance from the observer. (*Experiment and ITQ 11*)

ITQ ANSWERS AND COMMENTS

ITQ 1 8.64×10^4 seconds.

There are 60 seconds in a minute, 60 minutes in one hour, and 24 hours in one solar day. Therefore, there are:

$(60 \times 60 \times 24)$ seconds in one day,

i.e. 86 400 seconds in one day

or 8.64×10^4 seconds in one day

ITQ 2 10^6 seconds.

One week contains $7 \times 8.64 \times 10^4$ seconds $= 6.048 \times 10^5$ seconds.

To within an order of magnitude, we would say that there are 10^6 seconds in one week.

You should really have been able to write down this order of magnitude answer without having to do the full calculation. After all, if you only want the answer to within an order of magnitude, why do the calculation exactly? Because 7 multiplied by 8.64 is nearer to 100 than to ten, then, in a rough calculation, 10^4 must be increased by two orders of magnitude to 10^6.

ITQ 3 24 000 miles.

360° would correspond to $360 \times (500/7.5)$ miles, that is 24 000 miles around the circumference. But remember that an angle of 360° is defined to be the angle of a complete circle. So the distance round the circle corresponding to 360° must be the complete circumference. Hence the circumference of the Earth is 24 000 miles.

ITQ 4 3 820 miles.

According to Equation 5,

$$C = 2\pi R$$

Therefore

$$R = \frac{C}{2\pi}$$

Hence, putting in values for C and π,

$$R = \frac{24\,000}{2 \times 3.14} = 3\,820 \text{ miles}$$

(rounded-off to an accuracy consistent with the data)

If you are not happy with the rearranging of Equation 5 employed in this ITQ, you should refer to Into Science, *Module 8.*

ITQ 5 6 150 km.

Eratosthenes' value for the radius of the Earth was 3 820 miles. Since 1 mile is 1.61 km, 3 820 miles is $(3\,820 \times 1.61)$ km $= 6\,150$ km.

You may be interested to know that nowadays the generally accepted value for the radius of the Earth is about 6 380 km. The difference between this value and Eratosthenes's value is $(6\,380 - 6\,150)$ km $= 230$ km. As a fraction of the accepted value, this difference is 230 km/6 380 km ≈ 0.036. Expressed as a percentage, this difference is $0.036 \times 100\% = 3.6\%$.

ITQ 6 $\pi/2$ radians.

A right angle is 90°; and 90° is one-quarter of 360°. Hence, in radians, a right angle must be equal to one quarter of 2π radians, that is:

$$90° = 2\pi/4 \text{ radians}$$
$$= \pi/2 \text{ radians}$$

ITQ 7 We know that

$$2\pi \text{ radians} = 360°$$

Therefore

$$1 \text{ radian} = \left(\frac{360}{2\pi}\right)^{\circ}$$

$$\approx \left(\frac{360}{2 \times 3.14}\right)^{\circ} \approx 57.3°$$

ITQ 8 Equation 6 says that:

arc = radius × angle in radians

or angle = arc/radius (24)

The arc length has dimensions of length; the radius also has dimensions of length. Hence, the dimensions on the right-hand side of Equation 24 must be (length/length). That is, the right-hand side of this equation is dimensionless. If the dimensions of both sides of the equation are to balance, then the units of angle must also be dimensionless.

The radian is a dimensionless unit.

The same is true of any other unit of angular measure—the degree, for example, is also dimensionless. This is because an angle is really only a way of expressing a fraction of a rotation. So 3° really means 3/360 of a complete rotation, and 2 radians really means $2/2\pi$ of a complete rotation.

ITQ 9 3 820 miles.

The arc AS (i.e. the distance along the circumference from A to S) is given by:

arc AS $= R_E\,\theta$ (from Equation 6)

or equivalently

$$R_E\,\theta = \text{arc AS}$$

Dividing both sides of this equation by θ gives an expression for R_E,

i.e. $R_E = (\text{arc AS})/\theta$ (25)

But θ must be in radians; so, since

$$360° = 2\pi \text{ radians}$$

$$1° = \frac{2\pi}{360} \text{ radians}$$

and $\theta = 7.5° = \dfrac{7.5 \times 2\pi}{360}$ radians

You don't need to work this out yet. Instead, just replace θ (in Equation 25) *by this complete fraction*—after all, they are equal. Thus

$$R_E = (\text{arc AS})/\theta \qquad (25)^*$$

$$= (500 \text{ miles})\bigg/\left(\frac{7.5 \times 2\pi}{360}\right) \text{radians}$$

$$= \frac{500 \times 360}{7.5 \times 2\pi} \text{ miles}$$

$$= 3\,820 \text{ miles}$$

(rounded-off to an accuracy consistent with the data)

This is exactly the same answer that you arrived at before.

If you are not happy with the way in which this fraction was rearranged, try some of the examples in Into Science, *Module 2.*

ITQ 10 13.1 cm.

Equation 9 says that for small angles:

$$\theta \text{ (in radians)} \approx \frac{\text{shadow length}}{\text{pole height}}$$

Eratosthenes' value of θ was 7.5°, or $(7.5 \times 2\pi)/360$ radians (see ITQ 9).

Substituting this value of θ into Equation 9, together with a pole height of 100 cm, gives:

$$\left(\frac{7.5 \times 2\pi}{360}\right) \text{radians} \approx \frac{\text{shadow length}}{100 \text{ cm}}$$

Multiplying both sides of this equation by 100 cm gives:

$$\left(\frac{7.5 \times 2\pi}{360}\right) \text{radians} \times 100 \text{ cm}$$
$$= \text{shadow length.}$$

that is,

$$\text{shadow length} \approx 13.1 \text{ cm}$$

ITQ 11 Probably the best thing to do in this particular case is to take the *average* of the upper and lower limit values. For example, if you estimated that the largest possible radius was 12.5 cm, and the smallest possible radius was 7.5 cm, then the average value would be

$$\frac{(12.5 + 7.5)}{2} \text{cm} = 10 \text{ cm}$$

and the best estimate would be

$$(10.0 \pm 2.5) \text{ cm}$$

where adding on the 2.5 cm gives the upper limit, and subtracting the 2.5 cm gives the lower limit. (This result would be read as: R_E equals 10.0 cm, plus or minus 2.5 cm.)

Your value for R_E will probably be different to the one given here, which has simply been used to illustrate the method. *Your own result is the one you should enter in the space below the ITQ.*

ITQ 12 Suppose you found the radius of the Moon in the photograph to be (4.0 ± 0.1) cm, and the radius of the Earth (10.0 ± 2.5) cm. (*Note that these are 'made-up' figures—probably nothing like the ones you actually got.*) Then the *average* value of the ratio of the Earth radius to the Moon radius is:

$$\frac{10.0 \text{ cm}}{4.0 \text{ cm}} = 2.5$$

Now, a rough (though, in effect, somewhat pessimistic) estimate of the uncertainties associated with the ratio R_E/R_M can be found as follows. The *largest* value for the ratio would be obtained if you were to use the upper limit for the Earth radius and the lower limit for the Moon radius; that is:

$$\text{maximum possible value of } R_E/R_M = \frac{12.5}{3.9}$$
$$= 3.2$$

Conversely, the *smallest* value for the ratio is obtained by using the lower limit for R_E and the upper limit for R_M; that is:

$$\text{minimum possible value of } R_E/R_M = \frac{7.5}{4.1}$$
$$= 1.8$$

So we can write:

$$R_E = (2.5 \pm 0.7)R_M.$$

Comment: Perhaps you would like to check that you would have got more or less the same result if you had used 4.0 cm for R_M in both cases (rather than using 3.9 cm and 4.1 cm respectively). The reason why the uncertainty in R_M has very little effect on the final ratio is that the fractional (or percentage) uncertainty in R_M is swamped by the ten times larger fractional (or percentage) uncertainty in R_E.

This point is worth remembering. Whenever you combine several results, all of which have possible uncertainties associated with them, then the percentage error in the combined result will always be larger than the percentage error in any of the individual results. But if the percentage error in one measurement is much bigger than the percentage error in any of the other measurements, then the latter uncertainties can usually be neglected.

ITQ 13 In the 'made-up' example worked out in ITQs 11 and 12, the Moon was approximately two-fifths the size of the Earth; that is:

$$R_M \approx \left(\frac{2 \times 6\,200}{5}\right) \text{km} \approx 2\,480 \text{ km}$$

Remember that these values are just examples. You should, of course, calculate the radius of the Moon using your own estimate of the ratio of the two radii. You should also estimate the upper and lower limits for this radius.

ITQ 14 1.7 cm (approx).

Remember that the equation

$$\theta_M = d/l$$

is only true if θ_M is measured in radians. Recall that

$$1^\circ = 2\pi/360 \text{ radians}$$

Therefore, if $l = 1$ metre, and $\theta_M = 1^\circ$, then

$$d = l\theta_M = 1 \times \frac{2\pi}{360} \text{ metres}$$

$$\approx 1.7 \times 10^{-2} \text{ metres}$$

Hence, the diameter of the object needed to eclipse the Moon from a distance of 1 metre is *less* than 1.7 cm, i.e. less than the size of a 1p coin.

ITQ 15 Remember that

$$\text{arc} = R\theta$$

only holds true if θ is in radians.

$$3^\circ = \frac{2\pi \times 3}{360} \text{ radians} = \frac{\pi}{60} \text{ radians}$$

Therefore, from Equation 16:

$$L_S = \frac{L_M}{(\pi/60)} \approx 20L_M$$

i.e. (according to Aristarchus) the Sun is 20 times further away from Earth than is the Moon. We now know that this ratio is about 20 times too small.

ITQ 16 If you have obtained sensible values for the distances and sizes asked for in Table 4, then you should find that:

$$L_S \approx 24\,000 \text{ Earth radii}$$

If you obtained a value for L_S between $18\,000R_E$ and $30\,000R_E$, you've done pretty well! Mind you, if you got a value *very* close to $24\,000R_E$, then you have either been very lucky, or you've been cheating!

ITQ 17 This is not an easy question to answer concisely. Perhaps the best clue is contained in the words 'a fixed relationship'. If there is a 'fixed relationship' between R and T (to take Kepler's problem as an example), then there is always only one way of calculating T if you know R (and vice versa). For example, suppose the 'fixed relationship' were $T = 2R$. Then for any value of R, no matter how big or how small R might be, T would always be found by multiplying R by two.

Now consider one specific value of R and its corresponding value of T. Suppose R is increased by a small amount. Obviously T will increase by a correspondingly small amount (related to R through the fixed relationship $T = 2R$).

Successive increases in R will produce corresponding increases in T, that is, as R changes smoothly, T also changes smoothly. So whenever there is a *fixed relationship* between two quantities, the curve showing this relationship will, in general, be a smooth curve.

No such restriction will apply if there is *no* fixed relationship between the two quantities. In this situation, when R is changed by a small amount, it is quite possible for T to take on any value at all. Consequently the points could be scattered in no discernible pattern across the graph paper.

ITQ 18 If you obtained a smooth curve for Figure 26 (and you should have done), you should be feeling fairly confident that there is some form of fixed relationship between R and T. If this is the case, then you can deduce pairs of values for R and T other than those corresponding to the six planets you have plotted.

In other words, you would be able to say that, if a planet had an orbital radius R of 3.0 Earth-orbital radii, then it must also have a period T of about 5.2 Earth years. How do you know that it is 5.2? Simply by reading from your graph that T value corresponding to an R value of 3.0 Earth-orbital radii. That is, read vertically up the graph at $R = 3.0$ Earth-orbital radii until you get to the curve, and then note the T value corresponding to this point on the curve. This is shown in Figure 31. This process of finding pairs of values on graph (here, $T = 5.2$ and $R = 3.0$) lying *between* two plotted values, is known as *interpolation*.

ITQ 19 The curve you have drawn should pass easily through the point corresponding to the Earth, which fits nicely between the points corresponding to Venus and Mars. In Kepler's terms, this tells you that the Earth is the same kind of 'object' as the other planets. Copernicus was right—there is nothing special about the Earth. The relationship between T and R, which you have found in graphic form in Figure 26, is a relationship that applies to all planets orbiting the Sun including the Earth.

TABLE 9 For ITQ 20.

	T (Earth years)	R (Earth-orbital radii)	T^2 (Earth years)^2	R^3 (Earth-orbital radii)^3	T^2/R^3 (Earth years)^2/(Earth-orbital radii)^3
Mercury	0.24	0.39	0.058	0.059	0.98
Venus	0.62	0.72	0.384	0.373	1.03
Earth	1.00	1.00	1.00	1.00	1.00
Mars	1.88	1.52	3.53	3.51	1.01
Jupiter	11.86	5.20	141	141	1.00
Saturn	29.46	9.54	868	868	1.00

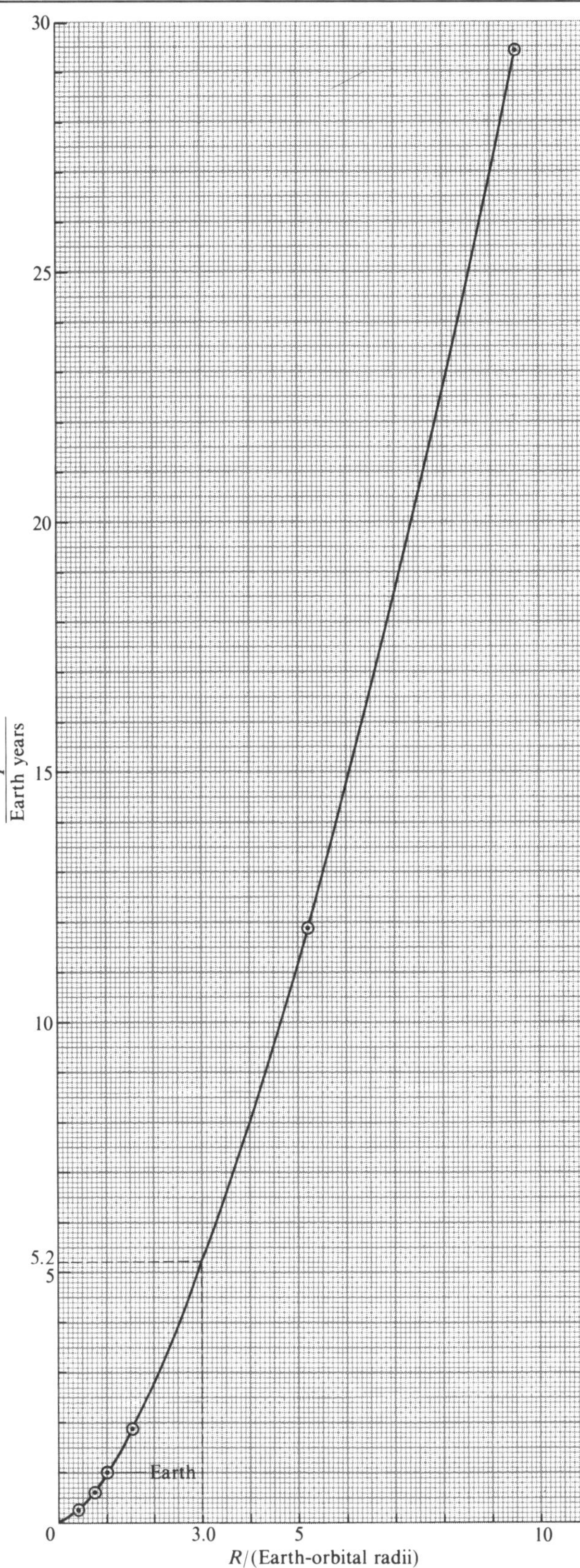

FIGURE 31 For ITQ 18.

ITQ 20 As you can see from the final column of Table 9, T^2 divided by R^3 comes to almost exactly unity for all the planets. The slight deviation is never more than a few per cent (a few parts in one hundred) and can be attributed to experimental uncertainty.

ITQ 21 83.74, 166.0 and 247.7 Earth years.

Kepler's third law says that:

$$T^2/R^3 = \text{constant} \qquad (20)^*$$

If T is measured in Earth years, and R in Earth-orbital radii, then the constant is exactly 1.00. Hence

$$T^2 = R^3 \qquad (22)^*$$

so

$$T = \sqrt{R^3} \qquad (26)$$

Alternatively, using the fractional index notation, we can write

$$T = (R^3)^{1/2} \qquad (27)$$

(A number to the power one-half means the square root of the number; a number to the power one-third means the cube root of the number; a number to the power $1/n$ means the nth root of the number.)

If you are not happy with this notation you should look at Into Science, *Modules 3 and 4.*

Using the radius values of Table 7 in Equation 27 gives:

Uranus: $T = (19.14^3)^{1/2}$

$\approx (7011.7)^{1/2}$

≈ 83.74 Earth years

(rounded-off to four digits)

Neptune: $T = (30.20^3)^{1/2}$

$\approx (27\,544)^{1/2}$

≈ 166.0 Earth years

(rounded-off to four digits)

Pluto: $T = (39.44^3)^{1/2}$

$\approx (61\,349)^{1/2}$

≈ 247.7 Earth years

These are the values of T predicted by Kepler's third law. The modern measurements for these values of T are 83.74, 165.95 and 247.69 Earth years, respectively. It looks as though Kepler was right! You should now feel confident enough to put aside your graph paper and use the formula instead.

ITQ 22 Trust the dimensional argument! The only way to make the dimensions on both sides of the equation balance, is to acknowledge that the constant in the equation must also have dimensions. Thus, writing the equation in terms of its dimensions:

$$(\text{time})^2 = (\text{dimensions of the constant}) \times (\text{length})^3$$

This equation balances if the dimensions of the constant are those of $(\text{time})^2/(\text{length})^3$, since the dimensions of the right-hand side can then be written as:

$$\frac{(\text{time})^2}{(\text{length})^3} \times (\text{length})^3 = (\text{time})^2$$

which is the same as on the left-hand side.

The *units* of the constant (as used in ITQ 20—see Table 9 in ITQ 20 answer) are (Earth years)^2/(Earth-orbital radii)^3. In SI units the constant would be in s^2/m^3.

ITQ 23 The value of the ratio T^2/R^3 is roughly the same for all four moons of Jupiter, about

$$2.4 \times 10^{-14}\,(\text{hours})^2/(\text{km})^3$$

Equation 20 is therefore satisfied: Kepler's third law does apply to the moons of Jupiter.

ACKNOWLEDGEMENTS

Grateful acknowledgement is made to the following for permission to reproduce Figures in this Unit:

Figure 1 English Heritage; *Figure 3* National Physics Laboratory, Crown Copyright; *Figure 13b* Yerkes Observatory; *Figure 18a* by courtesy of Patrick Moore; *Figure 18b* Royal Greenwich Laboratory; *Figures 20 and 27* Mansell; *Figures 22 and 23b* Ronan Picture Library; *Figure 28a* from J. J. Fahie, *Galileo his Life and Work*, Murray, 1903.

INDEX FOR UNIT 2